Robert Munro

Ancient Scottish Lake-Dwellings or Crannogs

With a supplementary chapter on remains of lake-dwellings in England

Robert Munro

Ancient Scottish Lake-Dwellings or Crannogs
With a supplementary chapter on remains of lake-dwellings in England

ISBN/EAN: 9783337233563

Printed in Europe, USA, Canada, Australia, Japan

Cover: Foto ©berggeist007 / pixelio.de

More available books at **www.hansebooks.com**

ANCIENT

SCOTTISH LAKE-DWELLINGS

OR CRANNOGS.

Edinburgh: Printed by Thomas and Archibald Constable

FOR

DAVID DOUGLAS.

LONDON . . . HAMILTON, ADAMS, AND CO.

CAMBRIDGE . . . MACMILLAN AND BOWES.

GLASGOW . . JAMES MACLEHOSE AND SONS.

LOCHSPOUTS 'LOOKING NORTHWARDS'

ANCIENT
SCOTTISH LAKE-DWELLINGS
OR CRANNOGS

WITH A SUPPLEMENTARY CHAPTER ON

REMAINS OF LAKE-DWELLINGS IN ENGLAND

BY

ROBERT MUNRO, M.A.

M.D., F.S.A. SCOT.

EDINBURGH: DAVID DOUGLAS

1882

PREFACE.

In publishing this work few prefatory remarks are required, beyond an explanation of the circumstances which led to its assuming the present form. The primary object contemplated was to place before general readers a record of some remarkable discoveries recently made in the south-west of Scotland, in a department of Archæology hitherto little known, and of which carefully prepared reports have already appeared in the second and third volumes of the Collections of the Ayrshire and Wigtownshire Archæological Association.

As it was at the instigation of R. W. Cochran-Patrick, Esq., M.P., that the explorations which led to these discoveries were originally undertaken, so it was also with him that the proposal to issue these reports in a handy volume originated. It occurred, however, to me, that, considering how little had been known of Scottish Lake-Dwellings in general, and that even this little was only accessible to the mem-

b

bers of a few learned Societies, it would be a more satisfactory undertaking to incorporate with the original reports, a *résumé* of the observations made by previous writers and explorers, so as to present to the public a complete compendium, as it were, of the whole subject.

The outcome of this idea is the volume now published, which, accordingly, aims at comprising all that is actually known of ancient *British* Lake-Dwellings up to the present time.

Such being its comprehensive scope, perhaps an apology ought to be here made for its many shortcomings; but this, I trust, will appear less necessary when I explain that it is the result of the occupation of such scraps of time as could be spared during the last two or three years from the active duties of a busy professional life.

Instead of attempting to interpret the references made to Lake-Dwellings by previous writers (many of which, though correct in point of fact, were little understood by the observers), in virtue of the additional knowledge derived from recent explorations, and giving the substance of their observations in my own language, I have thought it preferable to retain the exact words of the original narrators.   Hence my principal work, in the compilation of Chapter ii., was the selection from a mass of literature—chiefly old—of such portions as could be fairly construed to indicate the former existence of ancient lacustrine abodes in this country. The brevity of this portion of the work can be easily remedied by a perusal of the original sources from which my extracts have been taken, to all of which I have been careful in supplying the proper references.

The great services rendered to the science of Archæology by the numerous gentlemen who interested themselves in the various crannogs, and helped to bring the explorations to a successful termination, are faithfully acknowledged in the text where the respective investigations are described.

To R. W. Cochran-Patrick, Esq., LL.D., F.S.A., M.P., I am under the deepest obligations for valuable advice and assistance received in all stages of the researches —explorations, engravings, reports, etc.,—all being subject to his critical supervision. For the knowledge which he thus so freely and unselfishly placed at my disposal, as well as for much encouragement kindly given during the progress of the entire work, I now beg to express my warmest thanks.

To Joseph Anderson, Esq., LL.D., Keeper of the

National Museum of the Antiquaries of Scotland, I
am indebted for many hints regarding the character
of the relics, and for his kindness in reading the proof-
sheets of Chapter v.   The remarks made in this
chapter on the historical and traditional phenomena
associated with the Lake-Dwelling area in Scotland,
and on the supposed climatal changes since the prehistoric period, are intended as mere side-lights, and
for the benefit of general readers who may be curious
to know what (*i.e.* how little) the collateral sciences
have to do with the special branch of Archæology
treated of in this volume.

I have also to express my indebtedness to the
Ayrshire and Wigtownshire Archæological Association
for the use of all the woodcuts illustrating Chapters
iii. and iv., with the exception of Figs. 33 to 35, and
38 to 42, Fig. 54, Fig. 138, and Figs. 159 to 161;
and for permission to reprint the article of the late
Professor Rolleston on the Osseous Remains from the
Lochlee Crannog; that of Professor Bayley Balfour
on the Vegetal Remains from the same Crannog;
that of Mr. John Borland, F.C.S., F.R.M.S., on the
Analysis of Vivianite; that of Professor Cleland,
F.R.S., on the Osseous Remains from the Buston
Crannog; that of John Evans, Esq., D.C.L., F.R.S.,

F.S.A., on the Saxon Coin; and that of the Rev. George Wilson, C.M.S.A. Scot., on the investigation of Barhapple Crannog.

I am under similar obligations to the Society of Antiquaries of Scotland for the use of many of the woodcuts illustrating Chapter ii.

ROBERT MUNRO.

BRAEHEAD HOUSE, KILMARNOCK,
    May 1882.

CONTENTS.

CHAPTER I.

INTRODUCTORY.

CHAPTER II.

NOTICES, HISTORICAL AND DESCRIPTIVE, OF SCOTTISH LAKE-DWELLINGS PREVIOUS TO THE YEAR 1878.

CHAPTER III.

REPORT OF THE DISCOVERY AND EXAMINATION OF A CRANNOG AT LOCHLEE, TARBOLTON, AYRSHIRE.

CHAPTER IV.

SUBSEQUENT RESEARCHES AND DISCOVERIES AT FRIAR'S CARSE, LOCHSPOUTS, BARHAPPLE, AND BUSTON.

CHAPTER V.

GENERAL OBSERVATIONS ON THE CLASSIFICATION, GEOGRAPHICAL DISTRIBUTION, STRUCTURE, AND AGE OF ANCIENT SCOTTISH LAKE-DWELLINGS.

CHAPTER VI.

REMAINS OF LAKE-DWELLINGS IN ENGLAND.

APPENDIX.

LIST OF ILLUSTRATIONS.

PLATES.

CHAPTER I.

In searching back through the successive stages of human civilisation we arrive at a period when both written history and traditions fail. This prehistoric period, up to the commencement of the present century, was entirely lost in the thick veil of darkness which surrounded everything pertaining to the past history of the globe. As, however, the truths of geology became gradually formulated into a science, and men's minds got accustomed to apply the new methods of research to the elucidation of the origin and history of the human race, the sphere of prehistoric archæology became equally well defined. The group of phenomena with which it attempts to grapple occupies a sort of neutral territory or borderland between geology and history proper, from both of which, however, it receives large nutrient offshoots. The essential element which characterises the science of prehistoric archæology is an inductive process which depends on the clearness and precision with which the most primitive remains of human art and industries can be identified. But these remains, of whatever materials they may be composed, are liable to the destructive influences of time, and, sooner or later, they become obliterated by disintegration or decomposition. Few compound substances, even in the inorganic kingdom, resist this law, and as for the elaborate productions of the organic kingdom,

A

such as plants and animals, they are hardly ushered into
being when a counter process of decay begins, which ends
in reducing them to their simple constituents, so that, in a
short time, not a trace of their former existence remains. In
the midst of these ever-changing activities of life and death
which modern scientists have irrefutably shown to have been
continuously and progressively at work for countless ages,
it may be fairly asked—What is the nature of the evidence
by which antiquaries have so largely extended their field of
inquiry and propounded such startling opinions regarding
the origin and antiquity of our race? In their case the
evidence is due to exceptional circumstances which tend to
counteract or retard the gnawing tooth of time, and cheat, as
it were, Dame Nature out of her ordinary results. Thus, if
the handicraft products of reasoning man, or perishable
organisms, such as the bodies of animals, be accidentally
deposited in the mud of a sea, lake, or river, or suddenly
buried in the ruins of a city, or sunk in a bed of growing peat,
or become frozen up in a field of perpetual ice, these exceptional results are apt to follow. Hence, an object may be
preserved for centuries after its congeners, in ordinary circumstances, have crumbled into dust; or, if ultimately it
should become decomposed, a cast or mould may have been
previously formed by means of which, ages afterwards, an
intelligent observer will be enabled to determine its distin-
guishing characteristics. In arctic regions the carcasses of
animals known to have been extinct for hundreds of years
have been found imbedded in ice and so thoroughly preserved that their flesh was actually consumed by the dogs
of the present day ; and it is not a rare occurrence to find in
mossy bogs, such as those in Ireland, the bodies of human
beings, that have become accidentally buried in them centuries ago, completely mummified by the preservative influence of the matrix in which they have been entombed. In

short, these preservative qualities in nature are analogous to our artificial processes of pickling, embalming, or refrigerating, and had it not been for their occasional occurrence naturally, neither the science of geology nor that of prehistoric archæology would have much chance of being called into existence; nor could we now have any knowledge of the consecutive series of animals and plants that have inhabited this globe prior to the few centuries to which our historical records extend.

That these facts have failed to draw attention to lacustrine and other alluvial deposits as rich repositories of the remains of prehistoric man in Europe till about a quarter of a century ago, is more remarkable when we consider that ancient authors are not altogether silent on the habit prevailing among some races of erecting wooden abodes in lakes and marshes; that the Swiss lake villages, though singularly enough unnoticed by historians, were occupied as late as the Roman period; that frequent references have been made in the Irish annals to the stockaded islands, or as they are here called Crannoges, as existing in Ireland down to the Middle Ages; and that a similar custom is now found to be prevalent amongst some of the ruder races of mankind in various parts of the globe.

Hippocrates (*De Aeribus*, xxxvii.) speaks of the people in the Phasis, who live in the marshes, and have houses of timber and reeds constructed in the midst of the waters, to which they sail in single tree canoes. Herodotus (v. 16) also describes the dwellers upon the lake Prasias, whose huts were placed on platforms supported by tall piles in the midst of the lake, with a narrow bridge as an approach, and who, on one occasion, successfully resisted the military resources of a Persian army.

Villages composed of pile-dwellings are numerous along the shores of the Gulf of Maracaibo. "The positions chosen for their erections are near the mouths of rivers and in shallow waters. The piles on which they rest are driven deep into the oozy bottom, and so firmly do they hold that there is no shakiness of the loftily-perched dwelling perceptible, even when crowded with people. . . . Similar dwellings are found in other parts of South America, about the mouths of the Orinoco and the Amazons. They are the invention, not exactly of savages, but of tribes of men in a very primitive stage of culture."—(*Illustrated Travels*, vol. ii. pp. 19-21.)

Captain Cameron describes three villages built on piles in Lake Mohrya in Central Africa, and in his book of travels gives two sketches of these interesting abodes.—(*Across Africa*, vol. ii. p. 63.)

Captain R. F. Burton—"Notes connected with Dahoman" —refers to a tribe called Iso, who "have built their huts upon tall poles about a mile distant from the shore."— (*Memoirs of the Anthropological Society*, vol. i. p. 311.)

Pile-dwellings have been observed along the coasts of New Guinea and Borneo and the creeks and harbours running into the Straits of Malacca. In looking over some photographs recently brought from these regions, I was struck with one which is a representation of lake-dwellings at Singapore. The houses appear to be erected on a series of tall piles, and between the flooring and the water there is a considerable space in which the boats are hung up.

Though a few incidental notices of ancient lake-dwellings in Scotland preceded, in point of time, analogous discoveries in other countries, their real significance appears to have been overlooked till public attention was directed to the Irish Crannoges and Swiss Pfahlbauten. It is therefore desirable, on attempting to give a sketch of the work done in Scotland in this department of archæology, to give here a short account of these Irish and Continental discoveries, not only because they have been instrumental in opening up to Scottish antiquaries this wide field of research, the value of which as a storehouse of ancient relics is hardly yet realised, but because they enable us, by way of comparison, to point out some of the differences, as well as resemblances, of these ancient remains thus nominally associated under the common title of LAKE-DWELLINGS.

IRISH CRANNOGS.[1]

The historic references made to the Irish Crannogs are numerous, and extend over a long period, from the middle of the eighth down to the seventeenth century; but notwithstanding these, it was not till the year 1839 that their archæological importance became known. In this year Sir W. R. Wilde discovered and examined the crannog of Lagore, in the county of Meath, of which he has published an account in the first volume of the Proceedings of the Royal Irish Academy. After this other crannogs were discovered in rapid succession, and it became apparent that they existed very generally over the country. When Sir W. R. Wilde published his Catalogue of the Museum (Royal Irish Academy), in 1857, he states that no less than fortysix were known, and predicts that many others would be exposed to view as the drainage of the country advanced— a prediction which has been amply verified, because every succeeding year has seen an increase to their number.

---

[1] The word Crannoge, by which the artificial island fort was designated in the Irish Annals (modified by Drs. Robertson and Stuart into Crannog), is derived from the Gaelic *crann*, a mast or tree ; but as it is doubtful whether this etymology applies to the timber of which the island was constructed, or to the wooden huts erected over it, its use as a precise term to indicate the scope of this work would be equally doubtful. Hence I have preferred the word Lake-Dwelling.

According to this writer, crannogs " were not strictly speaking artificial islands, but cluans, small islets, or shallows of clay or marl, in those lakes which were probably dry in summer-time, but submerged in winter. These were enlarged and fortified by piles of oaken timber, and in some cases by stone-work. A few were approached by moles or causeways, but, generally speaking, they were completely insulated and only accessible by boat ; and it is notable that in almost every instance an ancient canoe was discovered in connection with the crannoge. Being thus insulated, they afforded secure places of retreat from the attacks of enemies, or were the fastnesses of predatory chiefs or robbers, to which might be conveyed the booty of a marauding excursion, or the product of a cattle raid."

A more recent explorer and writer on Irish crannogs, Mr. W. F. Wakeman, in a paper entitled " Observations on the principal Crannogs of Fermanagh," published in 1873,[1] goes on to say, after noticing their existence in eighteen different places in this county, and numbering no less than twenty-nine, " This glance is far from complete in its enumeration of the ' Lake Dwellings ' still remaining in this old territory, but it gives, I think, the principal examples. . . . . The Irish crannog, great or small, was simply an island, either altogether or in part artificial, strongly staked with piles of oak, pine, yew, alder, or other timber, encompassed by rows of palisading (the bases of which now usually remain), behind which the occupiers of the hold might defend themselves with advantage against assailants. Within the enclosure were usually one or more log-houses, which no doubt afforded shelter to the dwellers during the night-time, or whenever the state of the weather necessitated a retreat under cover."

<hr>

[1] *Journal of the Royal Historical and Archæological Association of Ireland*, vol. ii. p. 305.

None of the writers on Irish crannogs appear to have paid much attention to the structure of these islands, and beyond the mere statement that they were "stockaded," palisaded, or surrounded by one or more circles of piles, they give no explanation of the attachments and proper function of the surrounding piles. These are generally described as having been driven into the muddy bottom of the lake, and the most essential part of the mechanism of construction, viz., the horizontal mortised beams, has been only incidentally noticed. Though the purpose of these horizontal beams does not appear to have been understood, it is of importance to observe that their existence has not been entirely overlooked, as will be seen from the following quotations.

In his description of the crannog at Lagore near Dunshaughlin, Sir W. R. Wilde says : "The circumference of the circle was formed by upright posts of black oak, measuring from 6 to 8 feet in height; these were mortised into beams of a similar material laid flat upon the marl and sand beneath the bog, and nearly 16 feet below the present surface. The upright posts were held together by connecting crossbeams, and (said to be) fastened by large iron nails; parts of a second upper tier of posts were likewise found resting on the lower ones. The space thus enclosed was divided into separate compartments by septa or divisions that intersected each other in different directions; these were also formed of oaken beams in a state of good preservation, joined together with greater accuracy than the former, and in some cases having their sides grooved or rabbeted to admit large panels, driven down between them."[1]

Dr. Reeves, writing about a crannog in the county of Antrim, says : "These piles were from 17 to 20 feet long, and from 6 to 8 inches thick, driven into the bed of the

[1] *Proceedings of the Royal Irish Academy*, vol. i. p. 425.

lough, and projecting above this bed about 5 or 6 feet. They were bound together at the top by horizontal oak beams, into which they were mortised, and secured in the mortise by stout wooden pegs." [1]

My next quotation is from a paper by G. H. Kinahan, Esq., of the Geological Survey of Ireland, on Crannogs in Lough Rea : " A little N.W. of the double row, in the old working, there is a part of a circle of piles ; and in another, a row of piles running nearly E. and W. Mr. Hemsworth of Danesfort, who spent many of his younger days boating on the lake, and knows every part of it, informs me that on the upper end of some of the upright piles there were the marks of where horizontal beams were mortised on them. These seemed now to have disappeared, as I did not remark them." [2]

Mr. Wakeman, to whose writings I have already referred, writes as follows : " It would appear that, in some instances at least, their spike-like tops were anciently mortised into holes cut for their reception in beams of oak, which were laid horizontally. Just one such beam we found undisturbed, resting on the vertical spike *in situ*. A respectable elderly man, named Coulter, who resides not far from the lough (Ballydoolough) informed me that he well recollected to have seen many of these horizontal timbers resting upon the stakes or piles. They were hardly ever uncovered, but were distinctly visible a few inches below the surface of the water. This I believe to be a feature in the construction of crannogs but seldom remarked." [3]

As indications of the social economy and industries of the occupiers of these crannogs, there were found many

---

[1] *Proceedings of the Royal Irish Academy*, vol. vii. p. 155.

[2] *Ibid.* vol. viii. p. 417.

[3] *Journal of the Royal Historical and Archæological Association of Ireland*, vol. i. p. 362.

articles made of stone, bone, wood, bronze, and iron, such as swords, knives, spears, javelins, dagger-blades, sharpening-stones, querns, beads, pins, brooches, combs, horse-trappings, shears, chains, axes, pots, bowls, etc., and within the last few years, according to Mr. Wakeman,[1] many fragments of pottery, of a similar character to the fictile ware used for mortuary purposes in the prehistoric and pagan period, have also been found on some of them.

Many of these relics were deposited in the Museum of the Royal Irish Academy, but it is to be regretted that, owing partly to the system of classification now adopted in this Museum, by which articles are grouped together on the principle of resemblance, few of them can be identified or separated from the general collection, so that, except some articles thrown loosely into a drawer, and labelled as having been found in the crannogs of Dunshaughlin, Ballinderry, and Strokestown, no special or representative collection of crannog-remains now exists in Ireland. Several ancient canoes are well preserved in the lower portion of the Museum. Some have square-cut sterns, others have both ends pointed, some have cross bands, like ribs, left in the solid oak at regular intervals, as if to strengthen the sides of the vessel, while others are uniformly scooped out without any raised ridges. They vary much in size and shape. The largest is thus referred to in the small handbook to the Museum :—

"Down the centre of the room extends the largest known canoe, formed of a single tree. The remains measure 42 feet in length, and the canoe was probably 45 feet long, by from 4 to 5 feet wide, in its original state. It was recovered from the bottom of Loch Owel, in West Meath, and cut into eight sections for purposes of transport. There is a curious arrangement of apertures in the bottom, apparently

---

[1] *Journal of the Royal Historical and Archæological Association of Ireland*, vol. i. p. 383.

to receive the ends of uprights supporting an elevating deck."

On the antiquity of the Irish crannogs, Sir W. R. Wilde writes as follows:—"Certainly the evidences derived from the antiquities found in ours, and which are chiefly of iron, refer them to a much later period than the Swiss; while we do not find any flint arrows or stone celts, and but very few bronze weapons, in our crannogs. Moreover, we have positive documentary evidence of the occupation of many of these fortresses in the time of Elizabeth, and some even later."—(*Proceedings of Royal Irish Academy*, vol. vii. p. 152.) Subsequent researches, however, have shown that all the desiderated articles above mentioned have been found on crannogs. For instance, amongst the remains described by Mr. Shirley, from the crannogs in MacMahon's country, are stone celts, an arrow-head of flint, two arrow-heads of bronze, three looped bronze celts, bronze knives, etc.;[1] and G. H. Kinahan, Esq., M.R.I.A., thus concludes a short notice on Irish Lake-Dwellings, contributed to Keller's book (2d edit. p. 654):—"Of the time when the crannōgs were first built there is no known record, but that they must have been inhabited at an early period is evident, as antiquities belonging to the stone age are found in them. Some were in use up to modern times, Crannough Macknavin, County Galway, having been destroyed in A.D. 1610, by the English, while Ballynahuish Castle was inhabited fifty years ago. Some crannōgs seem to have been continuously occupied until they were finally abandoned, while others were deserted for longer or shorter periods. In Shore Island, Lough Rea, County Galway, there is a lacustrine accumulation over 3 feet thick, marking the time that elapsed between two occupations.

"In Wakefield's Island, A.D. 1812, attention was directed

---

[1] *Arch. Journal*, vol. iii. p. 47.

to a crannōg in Lough Nahincb, County Tipperary; but to the late Sir W. R. Wilde, M.D., is due the credit of bringing these structures prominently under public notice. This observer records forty-six crannōgs (*Catalogue, Royal Irish Academy*, vol. i. p. 220 *et seq.*), but since then twice as many have been recorded, most of which are described in the publications of the Royal Irish Academy, or kindred Societies; but a systematic classification of the crannōgs has yet to be made."

CONTINENTAL LAKE DWELLINGS.

Soon after the discovery of the Irish crannogs, the attention of archæologists was directed to remains of lake-dwellings in Switzerland. It appears that during the winter of 1853-4 the inhabitants of Ober Meilen, near Zürich, took advantage of the low state of the water in the lake to recover portions of the land, which they enclosed with walls, and filled in the space with mud. When the workmen began to excavate, they came upon heads of wooden piles, stone celts, stags' horns, and various kinds of implements. The late Dr. Ferdinand Keller, President of the Antiquarian Society at Zürich, hearing of the discovery, took up the matter with much energy, and after careful investigation of the remains at Ober Meilen, came to the conclusion that the piles had supported a platform, that on this platform huts had been erected, and that, after being inhabited for many centuries, the whole wooden structure had been destroyed by fire. Dr. Keller called these structures pile-buildings (Keltische Pfahlbauten), but they are more commonly known in this country as Lake-Dwellings (*habitations lacustres*). The discovery at Zürich was almost immediately followed by the discovery of similar structures in the other Swiss lakes. Owing to the vast system of drainage carried on since, there has been a great increase to their number, so that, at the present time, it is well ascertained that there was scarcely a sheltered bay in any of the lakes of Switzerland and neighbouring countries but contained a lake village. The most common plan adopted by the constructors of these ancient dwellings was to drive numerous piles of wood, sharpened sometimes by fire, sometimes by stone celts, or, in later times, by metal tools, into the mud near the shore of a lake; cross-beams were then laid over the tops of these piles, and fastened to them either by mortises or pins of wood, so as to form a platform. In certain cases the interstices between the upright piles were filled with large stones, so as to keep them firmer.

It appears also that the stones were brought in canoes and thrown down after the piles were driven in, in proof of which, a canoe, loaded with stones, was found in the Lake of Bienne, which had sunk to the bottom. Sometimes, when the mud was very soft, the upright piles were found to have been mortised into split oak-trees, lying flat at the bottom of the lake. Other erections were made by layers of sticks laid horizontally, one above the other, till they projected above the surface of the water, and thus presented a somewhat solid foundation for the platform. Upright piles here and there penetrated the mass, but rather served the purpose of keeping it together than of giving any support to the platform. These are called fascine-dwellings, and occur chiefly in the smaller lakes, and belong, for the most part, to the stone age.

The regular pile-buildings are far more numerous than the fascine-dwellings, but, notwithstanding the simplicity of structure of the latter, they do not appear to be older than the former, and it is a matter of observation that the civilisation of the fascine-dwellers corresponds with that of the inhabitants of other settlements of the stone age—in fact no difference has been observed between the earliest and the latest dwellings, except that the latter, as the result of improved tools, were found in deeper water.

The structural resemblance between the fascine-dwellings on the Continent, the Irish crannogs, and (as it will be afterwards seen) the Scottish lake-dwellings, is so striking, that the following, taken from Keller's book (2d edition, p. 597), is worth recording :—

" As the Lake of Fuschl is so near the Mondsee (Austria), it may be included in this notice; and it is somewhat singular, that here are found decided proofs of a 'fascine' lake-dwelling, in many respects similar to several found in Switzerland. This little lake and its banks are rich in fish and game. On the west side of the hill, where the former archiepiscopal hunting-lodge stood, there is a small bay with an island evidently made by human hands. It is nearly circular, about fifty paces in diameter, and is separated from the mainland by a narrow ditch or canal, now nearly filled up with moss and marsh plants. The island is covered first with a thick layer of peat moss and heather, beneath which lies a mass of branches, chiefly of the mountain pine and the dwarf birch. The island is very little raised above the water, and must have been very liable to be overflowed. The foundation appears to consist of boughs of pine-trees with their branches turned inwards. Small piles are driven in to keep them together, and, on the side of the lake, a number of stronger piles, or the remains of them, may be seen, amongst which lies a quantity of woody débris."

From the remains found on the sites of these lacustrine villages, it is inferred that their occupiers were acquainted with agriculture, and grew wheat and barley; that they had domesticated animals, such as cats, dogs, pigs, oxen, horses, sheep, and goats; that they used as food, besides the flesh of domesticated and wild animals, fish, milk, corn-meal boiled or baked, hazel nuts, plums, apples, pears, sloes, blackberries, and raspberries; that they were acquainted with the principles of social government and the division of labour; and that they manufactured cloth and ropes from bast and flax by means of looms, and the distaff and spindle. Their clothing consisted of skins of animals sometimes prepared into leather, as well as cloth plaited or woven from flax. Of the kind of huts or buildings erected over the platforms, little is known owing to their complete decay from exposure to sun and rain. They appear to have been rectangular in shape, and formed of wattle or hurdle-work of small branches, woven between the upright piles, and plastered over with clay. Each had a hearth formed of two or three large slabs overlying a bed of clay.

The earliest founders of these dwellings were, according to Keller, a branch of the Celtic population who came into Europe as a pastoral people, bringing with them, from the East, the most important domestic animals.

The absence of winter corn and hemp, most of the culinary vegetables, as well as the domestic fowl, which was unknown to the Greeks till about the time of Pericles, points to the period of their occupancy as a long way antecedent to the Christian era. Dr. Keller, one of the ablest authorities on this subject, has come to the conclusion that they were simply villages inhabited by a peaceful community, that they attained their greatest development about B.C. 1500, and that they finally ceased to be occupied about the commencement of the Christian era.

This wide chronological range embraces the three so-called ages of stone, bronze, and iron, but it appears that the settlements belonging exclusively to the stone age were more numerous and more widely distributed than those of the metallic period. Bronze age settlements were almost peculiar to western and central Switzerland, while the iron age is scarcely represented beyond the lakes of Bienne and Neuchâtel, so that it would appear that the lake-villages commenced to decrease in number towards the close of the former. Of the vast quantity and variety of relics found on their sites, illustrative of the culture and social organisation of their occupiers, it is impossible here to give even the barest description; but this is less necessary, as more detailed accounts are now easily accessible to general readers. After the voluminous and well illustrated work of the late Dr. Keller (as translated by Dr. Lee, 2d ed.), there is no epitome of the subject more worthy of perusal than chapter vi. of Sir John Lubbock's great work on Prehistoric Times, 4th edition.

CHAPTER II.

NOTICES, HISTORICAL AND DESCRIPTIVE, OF SCOTTISH LAKE-
DWELLINGS PREVIOUS TO THE YEAR 1878.

IT was not till these discoveries on the Continent had
attracted universal attention that Scottish archæologists
began to look for similar remains in this country. It was
then found that early historic references to island forts, and
some incidental notices of the exposure of buried islands
artificially formed of wood and stone, etc., during the drainage of lochs and marshes in the last, and early part of this,
century, had been entirely overlooked. The merit of correctly interpreting these remains, and bringing them systematically before antiquaries, belongs to Joseph Robertson, Esq.,
F.S.A. Scot., who read a paper on the subject to the Society
of Antiquaries of Scotland on the 14th December 1857,
entitled, " Notices of the Isle of the Loch of Banchory, the
Isle of Loch Canmor, and other Scottish examples of the
artificial or stockaded Islands, called Crannoges in Ireland,
and Keltischen Pfahlbauten in Switzerland."

This communication was not published in the Society's
Proceedings, the explanation of which will be found in the
following note, dated June 1866, which forms the introduction to a valuable article by Dr. Stuart, F.S.A. Scot., on
Scottish Crannogs :[1]—

" This paper was not printed in the Proceedings, in consequence of Mr. Robertson's desire to amplify his notices of these

[1] *Proceedings Soc. Antiq. Scot.* vol. vi. p. 114.

ancient remains. Other engagements having prevented him from carrying out his design, he recently placed his collections in my hands, with permission to add to my account of Scottish Crannogs anything from his notes which I might care to select. Of this permission I have gladly availed myself, and the passages introduced from Mr. Robertson's collection are acknowledged at the places where they occur.—J. S."

Mr. Robertson's paper, though not published, at once attracted attention, and stimulated so much further inquiry on the part of the members, that, at the very next meeting of the Society, another contribution on the subject was read by Mr. John Mackinlay, F.S.A. Scot., of which the following is an abstract. The paper is entitled " Notice of two ' Crannoges' or Palisaded Islands in Bute, with plans."[1]

DHU-LOCH, BUTE.

" The Crannoge of which I am now to give an account was discovered by me in the summer of 1812, and is thus described in a letter, dated 13th February 1813, which I wrote to the late James Knox, Esq. of Glasgow, who immediately sent it to his friend, George Chalmers, Esq., author of *Caledonia* :—' There is a small mossy lake, called Dhu-Loch, situated in a narrow valley in the middle of that strong tract of hill-ground extending from the Dun-hill of Barone to Ardscalpsie Point, to which valley, it is said, the inhabitants of Bute were wont to drive their cattle in times of danger. I remember, when a schoolboy, to have heard that there were the remains of some ancient building in that lake, which were visible when the water was low ; and happening to be in that part of the island last summer, I went to search for it. I found a low green islet about twenty yards long, which was connected with the shore, owing to the lowness of the water, after a continuance of

---

[1] *Proceedings Soc. Antiq. Scot.* vol. iii. p. 43.

dry weather.   Not seeing any vestiges of stone foundations, I was turning away, when I observed ranges of oak piles, and on examination it appeared that the edifice had been thus constructed.

"The walls were formed by double rows of piles, 4½ feet asunder, and the intermediate space appears to have been filled with beams of wood, some of which yet remain.   The bottom had been filled up to the surface of the water with moss or turf, and covered over with shingle, or quarry rubbish, to form a floor.   The ground-plan was a triangle, with one point towards the shore, to which it had been connected by a bridge or stage, some of the piles of which are still to be traced."

Mr. Chalmers, in his letter to Mr. Knox of 26th April 1813, relative to the above communication, says :—"It goes directly to illustrate some of the obscurest antiquities of Scotland—I mean the wooden castles—which belong to the Scottish period when stone and lime were not much used in building.   I will make proper use of this discovery of Mr. Mackinlay."

On revisiting this island in 1826, Mr. Mackinlay observed "an extension of the fort on the south-east corner, formed by small piles and a frame-work of timbers laid across each other, in the manner of a raft."

LOCH QUIEN, BUTESHIRE.

There was another insular fort in Loch Quien, which Mr. Mackinlay describes as a crannog ; but not being able to get on the islet, his measurements are conjectural, and need not be further referred to.   He then states that two rows of piles extended obliquely to the shore of the lake, between which the ground was covered with flat stones, "not raised like a causeway, but rather seeming to have been used as stepping-stones."

DR. ROBERTSON'S VIEWS ON SCOTTISH CRANNOGS.

Before resuming the chronological sequence of further discoveries, it becomes a matter of duty, on historical grounds, to refer more particularly to Mr. Robertson's views, notwithstanding that it is almost entirely to Dr. Stuart's elaborate paper, published some nine years later, that we are now indebted for any detailed record of his investigations. At the same time I shall take the opportunity of giving a few extracts of the incidental notices of artificial islands culled from other sources.

In the excellent article on Crannoges in *Chambers's Encyclopædia* (written, I believe, by Mr. Robertson), the following epitome of his opinions and researches is given :— " Hitherto, archæologists knew of lake-dwellings as existing only in Ireland and Switzerland; but in 1857, Mr. Joseph Robertson read a paper to the Society of Scottish Antiquaries, proving that they were to be found in almost every province of Scotland. He not only ascertained the existence of about fifty examples, but was able to show from records that they were known in Scotland by the same name[1] of Crannoges, which they received in Ireland. The resemblance between the Scottish and Irish types seems, indeed,

---

[1] "Instructions to Andro bischop of the Yllis, Andro lord Steuart of Vchiltrie, and James lord of Bewlie, comptroller, conteining suche overturis and articles as they sall propone, to Angus M'Coneill of Dunnyvaig and Hector M'Clayne of Dowart for the obedyence of thame and thair clanis. 14 Aprilis 1608. . . . That the haill houssis of defence strongholdis and *cranokis* in the Yllis perteining to thame and their foirsaidis sal be delyverit to his Maiestie and sic as his Heynes sall appoint to ressave the same to be vsit at his Maiesty's pleasour. . . . That they sall forbeir the vse and weiring of all kynd of armour outwith thair houssis especiallie gunis bowis and twa handit swordis, except onlie ane handit swordis and targeis."—(*Regist. Secreti Concilii: Acta penes Marchiarum et Insularum Ordinem* 1608-1623, pp. 4, 5. Robertson's *Notes.*)

to be complete. Every variety of structure observed in the one country is to be found in the other, from the purely artificial island, framed of oak-beams, mortised together, to the natural island, artificially fortified or enlarged by girdles of oak-piles or ramparts of loose stones; from the island with a pier projecting from its side, to the island communicating with the mainland by a causeway. If there be any difference between the crannoges of the two countries, it is that the number of crannoges constructed altogether of stones is greater in Scotland than in Ireland—a difference which is readily explained by the difference in the physical circumstances of the two countries. Among the more remarkable of the Scotch crannoges is that in the loch of Forfar, which bears the name of St. Margaret, the queen of King Malcolm Canmore, who died in 1097. It is chiefly natural, but has been strengthened by piles and stones, and the care taken to preserve this artificial barrier is attested by a record of the year 1508.[1] Another crannoge—that of

---

[1] The island in the Loch of Forfar, known as Queen Margaret's Inch, was discovered to be artificial on the partial drainage of the loch in 1781. It is thus referred to in the *Old Stat. Account of Scot.*, vol. vi. p. 528 :—

"Before this loch was drained, and near the north side of it, there was an artificial island composed of large piles of oak and loose stones, with a stratum of earth above, on which are planted some aspen and sloe trees, supposed to have been a place of religious retirement for Queen Margaret. This now forms a very curious peninsula. The vestiges of a building, probably a place of worship, are still to be seen. . . . It appears that the loch has at some period surrounded the rising ground, called the Manor, and the adjacent hill, on which the Castle of Forfar stood ; which hill is not, as the authors of the *Encyclopædia Britannica* suppose, artificial, but a congestum of sand and fat clay, evidently disposed in various irregular strata by the hand of nature."

Dr. Stuart says : *—"The drought of 1864 brought to light a sort of causeway, leading from the west end of the island. It was traced for about 100 yards ; and it is supposed that it turned to the shore on one side, the popular belief being that it formed a way of escape in former times. As, however, it must have formerly been under a great depth of

* *Proceedings Soc. Antiq. Scot.* vol. vi. p. 125.

Lochindorb, in Moray—was visited by King Edward I. of England in 1303, about which time it was fortified by a castle of such mark, that in 1336 King Edward III. of England led an army to its relief through the mountain passes of Athol and Badenoch.[1] A third crannoge—that of Loch Cannor or Kinord, in Aberdeenshire—appears in history in 1335, had King James IV. for its guest in 1506, and continued to be a place of strength until 1648, when the estates of Parliament ordered its fortifications to be destroyed. It has an area of about an acre, and owes little or nothing to art beyond a rampart of stones and a row of piles. In the same lake there is another and much smaller crannoge, which is wholly artificial.[2] Forty years after the dismantling

water, it seems doubtful for what purpose it may have been designed." (For historical notices of this Inch, see *Proc. Soc. Antiq. Scot.* vol. vi. p. 310.) Subsequent excavations (autumn of 1868) prove that St. Margaret's Inch is " the highest part of a narrow ridge of natural gravel which runs out into the loch, and the so-called causeway is the continuation of this ridge as it dips into the deep water." Dr. Stuart, who was present during the operations, remarks that the results obtained "afford another instance of the little reliance which can be placed on the descriptions of early remains given by the observers of last century, so far as relates to details." This is in allusion to Dr. Jamieson (*Archæologia Scotica*, vol. ii. p. 14) and others, who describe the Inch as being wholly artificial. (For Dr. Stuart's report, see *Proc. Soc. Antiq. Scot.* vol. x. p. 31.)

[1] "The only antiquity in the parish is the fortalix at Lochindorb, where a thick wall of mason-work (20 feet high even at this period, and supposed to have been much higher) surrounds an acre of land within the loch, with watch-towers at every corner, all entire. . . . Great rafts or planks of oak, by the beating of the waters against the old walls, occasionally make their appearance ; which confirms an opinion entertained of this place, that it had been a national business, originally built on an artificial island."—(*Old Stat. Account*, vol. vii. p. 259.)

[2] Before the level of Loch Canmor was reduced in 1858, it appears that it contained four islands—only one of which was found to be artificial. Of the three natural islands, the largest has an area of about a Scotch acre, and is known as the Castle Island, because the traditional castle of Malcolm Canmore was placed on it. It is supposed, from the occasional fishing up of great oak beams between it and the shore, that it was connected with the mainland by two projecting piers and a draw-

of the crannoge of Loch Cannor, the crannoge of Lochan-Eilean, in Strathspey, is spoken of as 'useful to the country

bridge. Fordun's expression "in turre sua de Canmore," and other historical references to the Isle of Loch Canmor apply therefore to this island with its castle, and not to the artificial island. The following extracts regarding the artificial island are from a paper on Loch Canmor, prepared by the Rev. James Wattie, Bellastraid, at the suggestion of Mr. Robertson, who intended to use it in his article on Scottish Crannogs :—

"The Prison Island is about the middle of the loch, and about 250 yards from its north shore. It is something of an oval shape. It is 25 yards long and 21 yards broad. It is evidently artificial, and seems to have been formed by oak piles driven into the loch, the space within the piling being filled up with stones, and crossed with horizonal beams or pieces of wood to keep all secure. The piles seem to have been driven or ranged in a rectangular form. They are quite distinct and apart from one another. The upright ones are generally round, though some of them have been splitted. The horizontal beams are mostly arms of trees, from 4 to 6 inches thick ; but there is one horizontal beam, squared evidently with an iron tool, about 8 inches on the side. There are not many horizontal beams now to be seen. I remember having seen more (the ends of trees) a good many years ago. My recollection of them is, that they had been splitted. There seems to have been upright piles on all sides of the island, but least distinct at the east end, and most numerous at the west. At the west end thirty upright piles are visible. On the south side, outside the regular row of piles, is a kind of out-fencing of upright and horizontal beams, seemingly for protection against the force of the water. At the west end there are two rectangular corners, and there may have been the same at the east end, though now overgrown with grass. Outside the piles is what may be called a rough, loose causewaying of stones sloping outwards into the water ; while inside is what may be called a heap of stones, arising, no doubt, from the putting into the water of whatever building had been on it. At the west end the piles stand 18 inches above the present level of the stones, and from 12 to 15 inches apart. They are 4 inches thick at the top, and 6 inches thick where they had been under water. Scarcely any of the upright piles are perpendicular; they slope to the north on the west side of the island, and to the west on the south side. Round the heap of stones now forming this island a clump of trees has sprung up. There is no appearance of a pier or jetty about the island, nor any mark of communication between it and the shore, or any of the other islands. The present depth of the loch near the island is 7 feet ; half-way between it and the Castle Island, 10 feet."

"On the 16th June 1859 there was fished up from the bottom of the

in times of troubles or wars, for the people put in their goods and children here, and it is easily defended.' Canoes (Fig. 1)

Fig. 1.—Canoe found in Loch Canmor.

hollowed out of the trunks of oaks have been found, as well beside the Scotch as beside the Irish crannoges.   Bronze

loch, near the north shore, opposite to the Prison Island, a canoe (Fig. 1), hollowed out of a single oak-tree, 22½ feet long, 3 feet 2 inches wide over the top at the stern, 2 feet 10 inches in the middle, and 2 feet 9 inches at 6 feet from the bow, which ended nearly in a point.   The edges are thin and sharp, the depth irregular—in one place 5 inches, the greatest 9 inches.   There are no seats nor rollocks or places for oars ; but there may have been seats along the sides, secured by pins through holes still in the bottom.   There are two rents in the bottom, alongside of each other, about 18 feet long each ; to remedy these, five bars across had been mortised into the bottom outside, from 22 to 27 inches long and 3 inches broad, except at the ends, where they were a kind of dovetailed, and 4 inches broad.   One of these bars still remains, and is of very neat workmanship, and neatly mortised in.   The other bars are lost, but their places are quite distinct.   They have been fastened with pins, for which there are five pairs of holes through the bottom of the canoe, at the opposite side, at a distance of from 18 to 20 inches, the bottom being flattish.   There are also five pairs of larger holes through the bottom, and also at the opposite sides, which may have been for fastening seats with pins along the sides of the canoe.   There are two bars mortised longitudinally into the bottom of the boat outside, above the seats before spoken of, 2½ inches broad, one at the stern 5 feet long, and the other beginning 5 feet from the stern, and extending 7½ feet towards the bow.   The canoe looks as it had been partly scooped out with fire.   The bottom is 2 feet 8 inches wide at the stern, and 28 inches wide at the middle.   The stern is 18 inches thick, and somewhat worn down at the top.

   " M'Pherson, the turner, says that twenty years ago a boat was taken up from the loch 26 feet long, sharp at both ends, otherwise coble-built, 8 feet broad in the bottom, which was flat, made of oak planks, overlapping one another, and lined under the overlapping with wool and tar." —(*Proc. Soc. Antiq. Scot.* vol. vi. pp. 167-171.)

vessels, apparently for kitchen purposes (Figs. 2, 3, 4, 5, 6), are also of frequent occurrence, but do not seem to be of a

Fig. 2.—(Height 10½ inches.) Found in Loch Canmor.

Fig. 3.—(Height 11 inches.)   Fig. 4.—(Height 9 inches.)
Found in the Loch of Banchory.

very ancient type. Deer's horns, boars' tusks, and the bones of domestic animals, have been discovered ; and in one instance a stone hammer, and in another what seem to be pieces for some such game as draughts or backgammon, have been dug up." (Fig. 7.)

Figs. 5 and 6. —Bronze Pots found in Loch of Banchory.

Fig. 7.—Found in Loch of Forfar (⅓).

THE ISLE OF THE LOCH OF BANCHORY.

" Before the recent drainage of the Loch of Leys—or the Loch of Banchory, as it was called of old—the loch covered about 140 acres, but, at some earlier date, had been four or five times as large. It had one small island, long known to be artificial, oval in shape, measuring nearly 200 feet in length by about 100 in breadth, elevated about 10 feet above the bottom of the loch, and distant about 100 yards from the nearest point of the mainland. What was discovered as to the structure of this islet will be best given in the words of the gentleman, of whose estate it is a part, Sir James Horn Burnett, of Crathes. In a communication which he made to this Society in January 1852, and which is printed in the first part of our Proceedings, he quotes from his diary of the 23d July 1850, as follows :—‘ Digging at the Loch of Leys renewed. Took out two oak trees laid along the bottom of the lake, one 5 feet in circumference and 9 feet long ; the other shorter. It is plain that the foundation of the island has been of oak and birch trees laid alternately, and filled up with earth and stones. The bark was quite fresh on the trees. The island is surrounded by oak piles, which now project 2 or 3 feet above ground. They have evidently been driven in to protect the island from the action of water.’ Below the surface were found the bones and antlers of a red deer of great size, kitchen vessels of bronze, a mill-stone (taking the place of the quern in the Irish crannogs), a small canoe, and a rude, flat-bottomed boat about 9 feet long, made, as in Ireland and Switzerland, from one piece of oak. Some of the

Fig. 8.—Isle of the Loch of Banchory. (General view of site.)

bronze vessels were sent to our Museum by Sir James Burnett, and are now on the table (Figs. 3 to 6). The general appearance of the island as it now is, since the bottom of the lake was turned into corn land, is represented by Fig. 8. The surface of the crannog was occupied by a strong substantial

building (Fig. 9). This has latterly been known by the name
of the Castle of Leys, and tradition, or conjecture, speaks
of it as a fortalice, from which the Wauchopes were driven
during the Bruces' wars, adding that it was the seat of the
Burnetts until the middle of the sixteenth century, when
they built the present Castle of Crathes. A grant of King
Robert I. to the ancestors of the Burnetts includes *lacum de
Banchory cum insula ejusdem.* The island again appears in
record in the year 1619, and 1654 and 1664, under the name

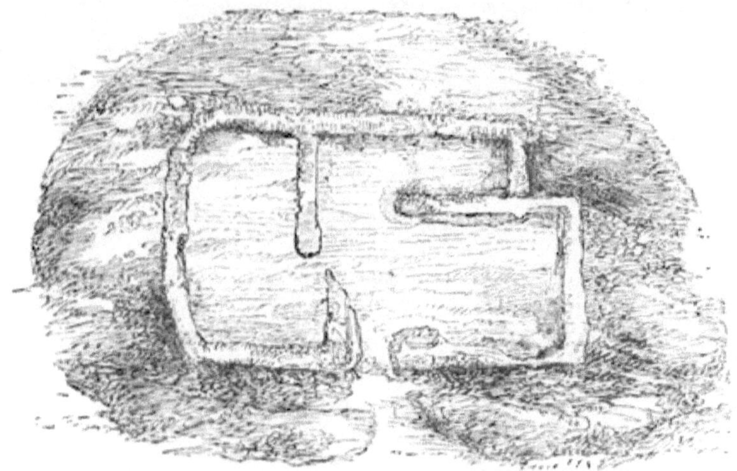

Fig. 9.— Isle of the Loch of Banchory.   (Surface of Crannog.)

of 'The Isle of the Loch of Banchory.' Banchory itself, I
may add, is a place of very ancient note. Here was the
grave of one of the earliest of our Christian missionaries, St.
Ternan, archbishop of the Picts, as he is called in the old
Service-Books of the Church, which add that he received
baptism from the hands of St. Palladius. Along with St.
Ternan's Head and St. Ternan's Bell, called the ' Ronnecht,'
there was preserved at Banchory, until the Reformation, a
still more precious relic, one of four volumes of the Gospel
which had belonged to him, with its case of metal wrought
with silver and gold."—(*Proc. Soc. Antiq. Scot.* vol. vi. p. 126.)

The following extracts regarding artificial islands incidentally observed in various parts of Scotland, brought to light chiefly in the course of drainage operations in search of marl or for the recovery of boggy land, may be now read with interest before resuming the narrative of more recent discoveries :—

LOCHRUTTON, KIRKCUDBRIGHTSHIRE.

" This loch is about a mile in length, and half a mile broad. In the middle of it there is a small island, about half a rood in extent, of a circular form. It seems to have been, at least in part, artificial. Over its whole surface there is a collection of large stones which have been founded on a frame of oak planks."—(*Old Stat. Account*, vol. ii. p. 37.)

LOCH KINDER, KIRKCUDBRIGHTSHIRE.

"In Loch Kinder there is an artificial mount of stones, rising 6 or 7 feet above the surface of the water, supposed to have been constructed for the purpose of securing the most valuable effects of the neighbouring families from the depredations of the borderers. The stones stand on a frame of large oaks, which is visible when the weather is clear and calm."—(*Old Stat. Account*, vol. ii. p. 139.)

CARLINGWARK LOCH, KIRKCUDBRIGHTSHIRE.

" When the water was let out of the Carlingwark Loch, in the year 1765, at the mouth of the drain next to the loch there was found a dam, or building of stone, moss, and clay, which appears to have been designed for deepening the loch. Besides this stone dam there was one of oak wood and earth, at the end of the town of Castle Douglas, now covered by the military road. About this place many horse-shoes were found sunk deep in the mud, of quite a different make from those now in use. Several very large stag-heads were got in the loch ; a large brass pan was also found in it. Near the south-west corner of the loch a brass *pugio* or dagger, 22 inches long, and plated with gold, was raised from the bottom in a bag of marl. Before it was drained there were two isles in the loch—the one near the north end, and the other near the south end of it. These isles were places of rest for large quantities of water-fowls of various kinds, which annually came and bred there ; even wild geese, it is said, have been sometimes known to breed on these isles. There was always a tradition in the parish that there had been a town in the loch which sunk, or was drowned; and that there were two churches or chapels—one on each of the large isles. . . . . The vestige or foundation of an iron forge was discovered on the south isle. Around it, likewise, there had been a stone building, or rampart; and from this isle to the opposite side, on the north-east, there is a road of stone secured by *piles of oak wood*, with an opening, supposed to have been for a drawbridge. In several places of the loch canoes were found which appear to have been hollowed, after the manner of the American savages, with fire. On a small isle, near the north end of the loch, there was found a large iron mallet or hammer stained on one end with blood. It is now in the hands of the Antiquarian Society at Edinburgh, and is supposed to have been an instrument used by the ancient Druids in killing their sacrifices. On several of the little isles in the loch were *large frames of black oak*, neatly joined. There are two small isles that have been evidently formed by strong piles of wood driven into the moss and marl, on which were placed large frames of black oak. The tops of

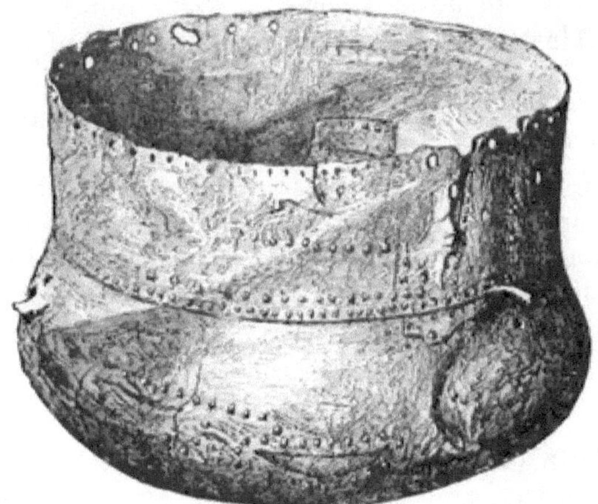

Fig. 10.—Caldron found in Carlingwark Loch.

these were fully 6 feet under water before the loch was drained. The design of these works is not at present known."—(*Old Stat. Account*, vol. viii. p. 304.)

A large bronze caldron, found by Mr. Samuel Gordon and J. T. Blackley while fishing near the Fir Island, and dredged out of the loch, contained a large number of iron and bronze implements (Fig. 10). When this caldron was raised, it was shining like gold. Mr. Gordon thinks it was left by Edward I., who had a camp on the Fir Island, and that it was deposited to prevent the Gallovidians getting any metal in their possession. A bronze sword was also found in this Loch, which is now in the National Museum, Scotland.—(*Proc. Soc. Antiq. Scot.* vol. vii. p. 7; and x. p. 286.)

Loch Spinie, Morayshire.

" Near it (the Bishop's Palace), where the water is deepest, a small artificial island emerged, upon clearing out the canal, of an oval form, about 60 by 16 paces, appearing to be composed of stones from the quarry, bound together by crooked branches of oak, and as if the earth, with which it was completed, had been wholly washed off during its submersion."—(*Old Stat. Account*, vol. x. p. 625.)

Loch of Boghall, Beith, Ayrshire.

" In the map of Cunninghame in Bleau's Atlas, published in 1654, there is laid down a piece of water called the Loch of Boghall. This loch belonged to the Monastery of Kilwinning, and was of old called Loch Brand. In the *Acta Dominorum Concilii* there is mention made of a case, 10th December 1482, at the instance of the Abbot and Convent of Kilwinning against, etc. etc., who were accused of the dangerous destruction and downcasting of the fosses and dikes of the Loch called Loch Brand. . . . The loch was drained about sixty years ago, when firm stakes of oak and elm were found in the soil, and which had been used for fixing the nets for fishing."—(*New Stat. Account*, vol. v. p. 580.)

Parish of Culter, Lanarkshire.

" In the midst of a morass, half a mile north-east from the farm of Nisbet, may be seen a very singular remnant of antiquity. A mound of an oval shape, called the Green Knowe, measuring about 30 yards by 40, rises about two or three feet above the surface of the surrounding bog. On penetrating into the elevating mass, it is found to consist of stones of all different kinds and sizes, which seem to have been tumbled promiscuously together without the least attempt at arrangement. Driven quite through this superincumbent mass are a great number of piles, sharpened at the point, about three feet long, made of oak of the hardest kind, retaining the marks of the hatchet, and still wonderfully fresh. A causeway of large stones connects this mound with the firm ground."—(*New Stat. Account*, vol. vi. p. 346.)

Loch Rannoch, Perthshire.

" There are two small islands situate in the upper end of the lake. The east and large one is wholly artificial, resting upon large beams of wood fixed to each other. This island was sometimes used as a place of safety in cases of emergency; at other times, as a place of confinement for such as rebelled against or offended the chief. To this retreat there is a road from a point on the south side—which road is always covered with 3 or 4 feet of water, is very narrow, and has a great depth on both sides of it."—(*New Stat. Account*, vol. x. p. 539.)

Parish of Croy, Inverness-shire.

" In draining a lake at the east-end of the parish, an artificial mound appeared within a few yards of the shore, about 60 feet in circumference, and 5 feet in height. It was formed of alternate strata of stones, earth, and oak; piles of oak being driven in the ground were kept strongly fixed by transverse beams of smaller size. Over these were round stones, and on the surface some inches of fine black mould. Some fragments of brass rings, pieces of potteries, and the bolt of a lock, of no ordinary size, were found on the mound.

" At about 100 yards' distance there is a circle of large piles of oak, driven deep in the earth, apparently the commencement of a second mound; but for what purpose they were intended it is impossible to conjecture. They could not be places of defence, as the one finished was so near the edge of the lake, and completely commanded by the opposite rising bank. While draining the lake by cutting a deep canal, oaks of gigantic size were found more than 20 feet below the surface, as sound as the day they were overwhelmed by water, sand, and gravel. At the same time a canoe of most beautiful workmanship was found, which some modern Goth has since cut down for mean and servile purposes."—(*New Stat. Account*, vol. xiv. p. 448.)

LOCHS OF KINELLAN AND ACHILTY, CONTIN, ROSS-SHIRE.

"In Lake Kinellan stands an artificial island, resting upon logs of oak, on which the family of Seaforth had at one period a house of strength. . . . There is still in Loch Achilty a small island, likewise supposed to be artificial. It belonged to Mac Lea Mor, *i.e.* Great Mac-Lea, who possessed at the same time a large extent of property in the parish; and who was wont, in seasons of danger, to retire to the island as a place of refuge from his enemies. The ruins of the buildings which he there occupied may still be traced."—(*New Stat. Account,* vol. xiv. p. 238.)

LOCH COT, PARISH OF TORPHICHEN.

"The loch lies at the foot of the southern slope of Bowden Hill, and is now drained. An old man who belonged to Dr. Duns's (New College, Edinburgh) congregation when he was at Torphichen, more than once described to him the appearance of the loch before it was drained—'its central island, and the big logs taken from it and burned.' Horns were also found in the loch, but were neglected, and have disappeared. Dr. Duns found part of a quern on an examination of the site; and on digging into a mound at a short distance eastward from the loch, he found an urn of rude type. To the south are the remains of a circular earthwork; and to the south-west, traces of what has been called a Roman camp; and to the south, a camp of peculiar form, noticed by Sibbald."—(Dr. Stuart's article, *Proc. Soc. Antiq. Scot.* vol. vi.)

CASTLE LOCH, LOCHMABEN.

In the Castle Loch of Lochmaben is a small artificial island now sunk several feet under the water, from which during dry weather on several occasions some of the oak mortised beams have been fished up.

LOCH LOCHY, INVERNESS-SHIRE.

Dr. Stuart quotes the following account of a crannog in Loch Lochy from Mr. Robertson's notes, extracted by the latter from a MS. in the Advocates' Library, written towards the end of the seventeenth century. "Ther was of ancient ane lord in Loquhaber, called my Lord Cumming, being a cruell and tyirrant superior to the inhabitants and ancient tenants of that countrie of Loquhaber. This lord builded ane iland or an house on the south-east head of Loghloghae; . . . and when summer is, certain yeares or dayes, one of the bigge timber jests, the quantitie of an ell thereof will be scin above the water. And sundrie men of the countrie were wont to goe and se that jest of timber which stands there as yett; and they say that a man's finger will cast it too and fro in the water, but fortie men cannot pull it up, because it lyeth in another jest below the water."— (*Proc. Soc. Antiq. Scot.* vol. vi. p. 160.)

Loch Lomond.

Regarding a crannog in Loch Lomond,[1] in the neighbourhood of a stone cashel on shore, from which large mortised joists were disjoined in 1714, and used by a gentleman in that country for building a house, see extracts from Dr. Robertson's notes, *Proc. Soc. Antiq. Scot.* vol. vi. p. 132.

Loch of the Clans, Nairnshire.

In 1863 a paper by Dr. Grigor of Nairn was read at the Society of Antiquaries of Scotland, in which the author describes a curious cairn with "oak beams and sticks cropping out in it," and surrounded by a ploughed field which was formerly part of the basin of the Loch of the Clans, from which I give the following extracts:—

"On getting down into the cairn, we found that all the wood in sight was chiefly the remains of rafters, and inclined upwards at about an angle of 25 degrees, so as to form an upright roof. These, however, had been broken across (as represented on the sketch, Plate I.), no doubt by their own

---

[1] In Maitland's *History of Scotland* I find the curious statement made that Boece states that in Loch Lomond there were fish without fins, waves without wind, and a floating island.—(*Boet. Scot. Reg. Descript.* fol. 7.)

partial decay and the superincumbent weight of stones. On further clearing and digging, we came upon four sides or walls, each about 3 feet in height, and making an irregular square. These were formed of trees of oak, comparatively sound, and about thirty years' probable growth. On the west side, there were seven trees piled horizontally, one above the other; the third from the ground had another alongside it. Seven trees also formed the east side. The north side was made up of a foundation of small boulders, then two horizontal trees, over which projected a few rafters, and then another tree. At the east end of this wall there was a mortised opening, in which, in all probability, an upright support had been placed. The south side had been, to all appearance, partially removed when that end of the cairn was carted off by the tenant-farmer, and only one tree at the bottom was seen. These sides are correctly represented in the accompanying sketches, and the scale renders it unnecessary to particularise measurements. The floor was the mud bottom of the old loch, and there were two small trees stretching from east to west, with the appearance of decayed brushwood throughout, and a boulder stone here and there. Not exactly in the centre, but nearer the south-east corner, lay a few boulders bearing marks of fire, and having portions of charcoal around them. This was all that could be seen as a hearth.

"Nothing of any interest was found in the way of clearance. There were portions of decomposed bones, a bit of pottery (evidently modern), the mouth-piece of a horn spoon, and a cockle shell, and these probably had fallen through the cairn.

"To all appearance the rafters started from the ground in three tiers, having different angles of inclination, though those of the roof seemed to have run up pretty much together near the ground. These were bound down by beams

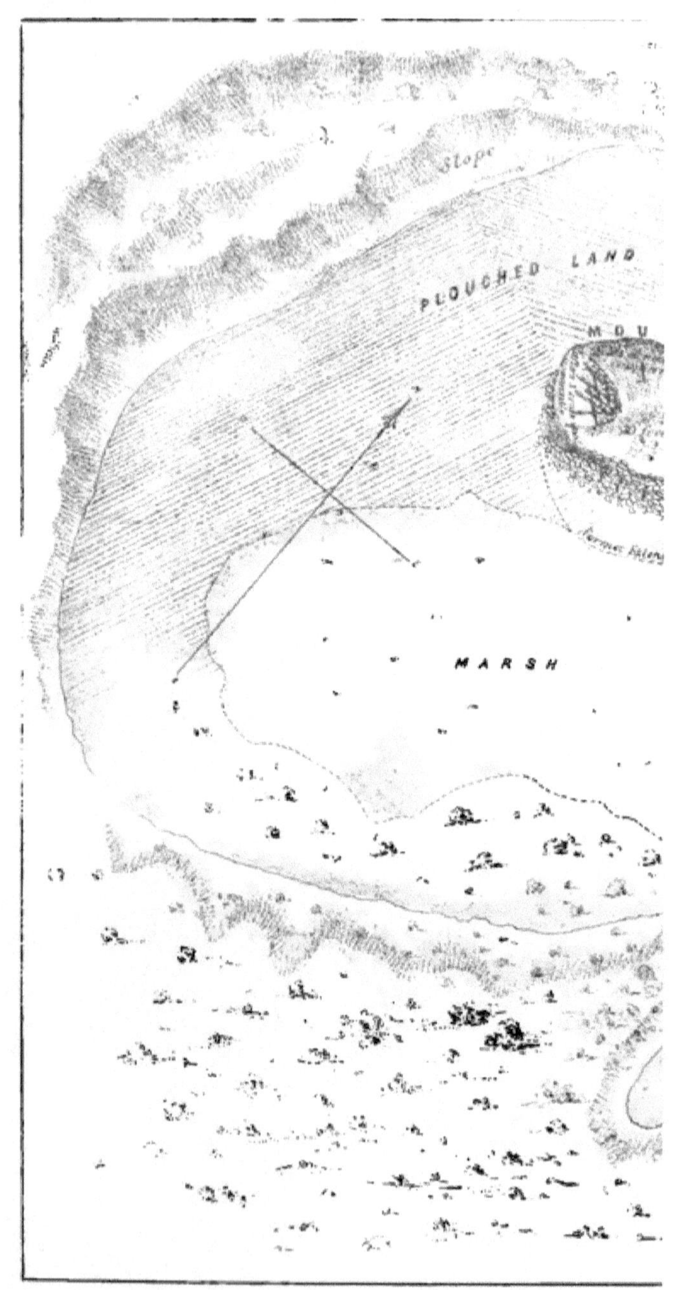

Slope
PLOUGHED LAND
MOU
MARSH

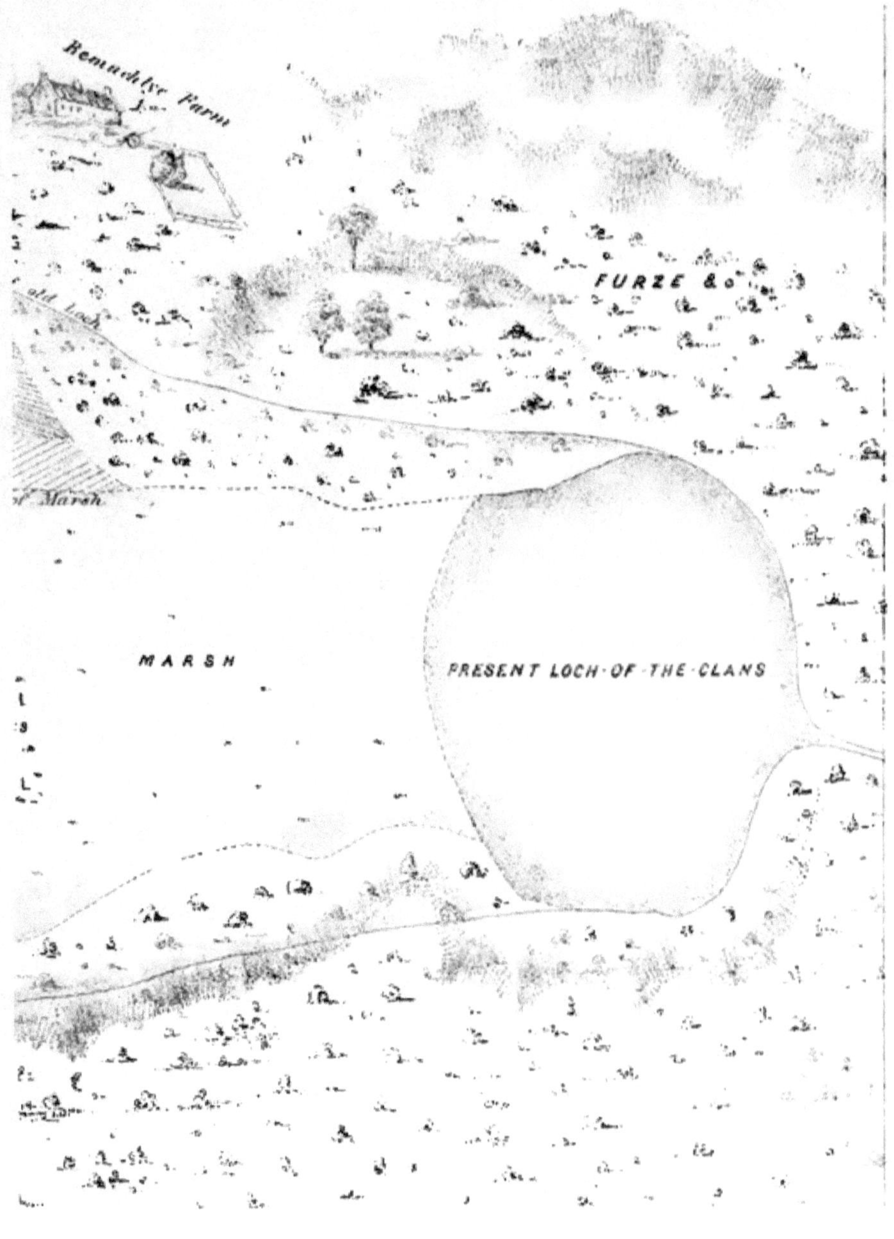

Remarkhie Farm
old Loch
of Marsh
FURZE &c
MARSH
PRESENT LOCH OF THE CLANS

crossing and recrossing in all directions, which imparted greater strength. Beyond two mortised openings no other mode of fastening could be seen."

The articles found in the course of the explorations of this cairn, or crannoge, were a portion of a small stone cup, two whetstones, and an iron axe-head, together with charcoal and burnt bones. In the vicinity there were found some flint arrow-heads and flakes, and, some years ago, a canoe.

" About 150 feet, in a south-easterly direction from this place, and in marshy ground, were found a great many pile-heads, covered with grass and vegetable matter; and after removing this covering they stood as shown in the accompanying plan and scale. This is no doubt the foundation of another crannoge or lake habitation. An area of 6 feet in the centre seemed, so far as I examined, to have been laid with large stones, and intersected with small trees and stakes. Beyond this space I observed no stones, only the mud of the lake, and a few bits of small trees. Three stones in the centre seemed marked by fire; and below those I turned over, and under water, there was a good deal of charcoal mixed with small bits of bone.

" In the neighbouring ' Loch of Flemington,' and covered with several feet of water, are to be seen, when the water is frozen over, similar remains of piles.

" In the east end of the small pond called ' Loch in Dunty,' about two miles in a westerly direction from that of Flemington, are to be observed three vestiges of piles about a foot above water; these, notwithstanding the evidence of a Highlander living close by, ' that the piles had been put into the loch in auld time, for the purpose of steeping the lint,' are, in my opinion, of the same description, day and generation, as those I have attempted to describe in the Loch of the Clans."—(*Proc. Soc. Antiq. Scot.* vol. v. pp. 116, 332.)

Loch of Sanquhar, Dumfries-shire.

In June 1863, Dr. Grierson of Thornhill announced, at a meeting of the Dumfries-shire and Galloway Natural History Society, that an ancient stockade had been found in a small loch near Sanquhar. He observed, " that about five weeks ago, a man drowned himself in a tarn about two miles north of Sanquhar. In order to recover the body, the water was drained off, when it was found that a small island in the middle of the loch or tarn was artificial, and had been constructed of stakes with stones between, and had been approached by a zigzag line of stepping-stones. It was thought that the loch might be altogether artificial, forming, as it were, a moat or fosse to the little fort."—(*Proceedings,* Session 1863-4, p. 12.)

During the summer of 1865, the members of the Society made an excursion to this loch for the purpose of examining the crannog, the result of which is described by the President (the late Sir W. Jardine, Bart.) in his annual address, and from which the following extracts are taken :[1]—

"This loch is of considerable depth, and now covers about 2 acres. At the north end of this there is a small island covered with a rank vegetation of grasses, carices, etc., mixed with a few plants of *Epilobium angustifolium,* and there are also a few stunted trees of Scotch fir and birch. At the north or north-east end there is a natural outlet from the loch through the moss, which could be easily deepened." . . . (This outlet was deepened previous to their visit, and the water drained off so as to facilitate the examination of the island.)

"When first seen, after the bottom was laid dry, a few upright piles were observed, and the curving narrow passage from the mainland appeared somewhat raised, and was hard below the immediate mud deposit, as if a sort of rough causeway had been formed; and when the water was at its height, or nearly level with the surface of the island, persons acquainted

[1] *Proceedings,* 8th December 1865.

with the turn or winding of the passage could wade to it. The base of the slope of the island was laid or strengthened with stones, some of considerable size, so placed as to protect the wooden structure. Round the island could be seen driven piles, to which were attached strong transverse beams, and upon making a cut 6 or 7 feet wide into the side of the island to ascertain its structure, we found a platform of about 4 feet in depth raised by transverse beams placed alternately across each other, and kept in position by driven piles. These last were generally self oak trees, but dressed and sharpened by a metal tool, some of them mortised at the heads where a transverse rail or beam could be fixed. The transverse beams, of various sizes, were chiefly of birch wood. . . . On the surface of the island there were some indications of buildings, but on examination these were found to be only the erection of curlers for fire, or the protection of their channel-stones when not in use. No remains of any kind were found on the island nor around it, but, except on the passage from the mainland, the mud was so deep and soft as to prevent effectual search. Neither have we any record of any other remains being found in or near the loch except the canoe already alluded to. It is formed out of a single oak tree, 16 feet in length by 3 feet broad at the widest part, at the prow only 1 foot 10 inches."

Loch Barean, Kirkcudbrightshire.

The following facts were communicated to the author by John J. Reid, Esq., F.S.A. Scot., Edinburgh :—

Loch Barean, situated in a mountainous patch in the parish of Colvend, is about five miles from Dalbeattie, and not far from the main road leading from this town to Colvend Manse. It is a deep peaty sort of a loch, very irregularly shaped, measuring about 1950 feet in length, with a breadth varying from 600 to 1150 feet, and is bounded on the east by a barren ridge of rock which runs along its margin. In 1865 the level of this loch was lowered by drainage, when a few stones, which used to become visible in dry summers, turned out to be an artificial island constructed of wooden beams. Shortly after exposure, it was visited by the late Sir W. Jardine, Dr. Stuart, Secretary to the Society of Antiquaries, and others, and found to be surrounded by a circle of oak piles enclosing a wooden flooring. "None of these piles were visible above the water. On this oak piling beams had been laid horizontally, some oak and some of fir still retaining the bark. The space within this piling was nearly circular in shape, and measured about 24 feet in diameter; but, outside the piling, and between it and the loch, there was an area, from 5 to 8 feet wide, filled with angular granite blocks to assist in protecting the wooden flooring."

Two metal "pots" were found on the island, of which only one now remains. It is of thin beaten bronze, flat-bottomed, with bulging sides and everted lip. Its dimensions are: height 5 inches, diameter of mouth $4\frac{3}{4}$ inches, do. of bottom $3\frac{3}{4}$ inches, and do. in middle $5\frac{3}{4}$ inches.[1]

CRANNOGS IN LOCH DOWALTON.

A more important discovery, made about the same time, was a group of artificial islands in Lake Dowalton, Wigtownshire, which were first described by Lord Lovaine, in a paper read to the British Association in 1863. Mr. John Stuart, Secretary to the Society of Antiquaries of Scotland, then took up the subject, and, owing to a greater drainage of the loch having been made in the interval, was enabled to re-examine the Dowalton islands under more favourable circumstances. The result of his labours was an elaborate paper to the Society, in which he gave a detailed account of the structure and relics of these crannogs, and also took the opportunity of incorporating into his article all the facts he could glean, so as to afford a basis for comparing the Scottish examples with those in other countries. I have taken ad-

---

[1] See also *Proceedings of Dumfries and Galloway N. H. Soc.* for 1865.

vantage of some of the contents of this paper on a previous occasion when discussing Mr. Robertson's investigations of Scottish crannogs. The following is the substance of Dr. Stuart's examination and report of the Dowalton group:[1]—

The late Loch of Dowalton was of an irregular form, about 1½ mile long, and about ¾ths of a mile in greatest breadth, and without any marked outfall for drainage. Sir William Maxwell effected this by making a cut, 25 feet deep, through the wall of whinstone and slate which closed it in at its south-eastern extremity. Dr. Stuart, who availed himself of Lord Lovaine's previous description of the island abodes that became visible on the drainage of the loch, describes them in order of succession, beginning at the west end :—

"The first is called Miller's Cairn, from its having been a mark of the levels, when the loch was drained by cuts for feeding neighbouring mills. One of these cuts is known to have been made at a remote period. It was still surrounded by water when the place was visited by Lord Percy in 1863. On approaching the cairn, the numerous rows of piles which surrounded it first attracted notice. These piles were formed of young oak-trees. Lying on the north-east side were mortised frames of beams of oak, like hurdles, and, below these, round trees laid horizontally. In some cases the vertical piles were mortised into horizontal bars. Below them were layers of hazel and birch branches, and under these were masses of fern, the whole mixed with large boulders, and penetrated by piles. Above all was a surface of stones and soil, which was several feet under water till the recent drainage took place. The hurdle frames were neatly mortised together, and were secured by pegs in the mortise holes.

"On one side of the island a round space of a few feet in size appeared, on which was a layer of white clay, browned

<hr>

[1] *Proceedings Soc. Antiq. Scot.* vol. vi. pp. 114 *et seq.*

and calcined as from the action of fire, and around it were bones of animals and ashes of wood. Below this was a layer of fern and another surface of clay, calcined as in the upper case. A small piece of bronze was found between the two layers. On the top another layer of fern was found, but the clay, and the slab which probably rested upon it, had been removed. There can be no doubt that this had been used as a hearth. Near this cairn a bronze pan was found. . . . Lines of piles, apparently to support a causeway, led from it to the shore.

"The next in order is the largest island. Lord Percy succeeded in reaching it in a boat in 1863. It appeared to him to be 3 feet below the level of the other islands, and, from several depressions on its surface, to have sunk. The progress of excavation was, however, soon checked by the oozing in of the water. On the south side of the island great pains had been taken to secure the structure; heavy slabs of oak, 5 feet long, 2 feet wide, and 2 inches thick, were laid one upon another in a sloping direction, bolted together by stakes inserted in mortises of 8 inches by 10 inches in size, and connected by square pieces of timber 3 feet 8 inches in length. The surface of the island was of stones, resting on a mass of compressed brushwood, below which were branches and stems of small trees, mostly hazel and birch, mingled with stones, apparently for compressing the moss. Below this were layers of brushwood, fern, and heather, intermingled with stones and soil, the whole resting on a bed of fern 3 or 4 feet in thickness. The mass was pinned together by piles driven into the bottom of the loch, some of which went through holes in the horizontal logs. I noticed some of these flat beams of great size and length (one of them 12 feet long), with *three mortise holes* in the length, 7 inches square. A thick plank of oak of about 6 feet in length had grooves on its two edges, as if for something to slide in. This island measured 23 yards across, and was surrounded by many rows of piles, some of which had the ends cut square over, as if by several strokes of a small hatchet. Vestiges of branches were observed interlaced in the beams of the hurdles. On the north-east side, and under the superstructure of the island (hurdles and planks), a canoe was found, made of a single tree of oak. It was 21 feet in length, 3 feet 10 inches across over all near the stern, which was square. Its depth at the stern was 17 inches, or, including the back-board which closed the stern, 20 inches. The stern was formed by a plank inserted in a groove on each side, with a back-board pegged on above it. The part containing the grooves was left very thick. There were two thole-pins on each side, inserted in squared holes in the solid, which was left to receive them, and wedged in with small bits of wood. One thwart of fir or willow remained. A plank or wash-board projecting a few inches over the edge, ran round the canoe. It rested on the top, and was fastened with pegs into the solid. . . .

"On one spot a few flat stones were placed as if for a hearth. The best saucepan was found between this island and the shore. A small circular brooch of bronze, four whet-stones, two iron hammers, and some lumps of iron slag, were found *on* the island. A third iron hammer was found near it.

"The original depth from the surface of the island to the bottom was probably from 6 to 7 feet; but the structure was much dilapidated before I saw it.

"Proceeding southward, we come to the island first examined by Lord Percy. It proved to be nearly circular, and to be about 13 yards in diameter. Its surface was raised about 5½ feet above the mud, and on each side of it were two patches of stone nearly touching it. On the north side lay a canoe of oak, between the two patches, and surrounded by piles, the heads just appearing above the surface of the mud. It was 24 feet long, 4 feet 2 inches broad in the middle, and 7 inches deep, the thickness of the bottom being 2 inches. Under the stones which covered the surface, teeth of swine and oxen were found. A trench was cut round the islet, and at the south end a small quantity of ashes was turned up, in which were teeth and burned bones, part of an armlet of glass covered with a yellow enamel, and a large broken bead of glass, together with a small metal ornament; two other pieces of a glass armlet, one striped blue and white, were also found on the surface. These objects were found on the outside of the islet, about 2 feet from the surface. On cutting into the islet itself, it proved to be wholly artificial, resting on the soft bottom of the loch, and in its composition exactly the same as the large island already described. The whole mass was pinned together by piles of oak and willow, some of them driven 2½ feet into the bottom of the loch. The islet was surrounded by an immense number of piles, extending to a distance of 20 yards around it; and masses of stone, which apparently were meant to act as breakwaters, were laid amongst them. On the sinking of the mud, a canoe was found between the islet and the northern shore. It was 18½ feet long, and 2 feet 7 inches wide. A block of wood cut to fill a hole, left probably by a rotten branch, was inserted in the side, 2 feet long, 7 inches wide, and 5½ inches thick, and was secured by pegs driven through the side; across the stern was cut a deep groove to admit a back-board; in both canoes a hole 2 inches in diameter was bored in the bottom.

"The next islet is about 60 yards from the last, and nearer to a rocky projection, on the south margin of the loch. It was examined by Lord Percy and was found to be smaller; the layers were not so distinctly marked, and some of the timbers inserted under the upper layer of brushwood were larger, and either split or cut to a face. A stake with two holes bored in it about the size of a finger, a thin piece of wood in which mortises had been cut, and a box, the interior of which was about 6 inches cube, with a ledge to receive the cover, very rudely cut out of a block of wood, were found.

"On the south-east side of the loch, near one of the little promontories, were several cairns surrounded by piles, of which the outline had mostly disappeared at the time of my visit. When they were first seen by Lord Percy, there were six structures of the same character as those already described, arranged in a semicircle. They were, however, much smaller than the others, and appeared to have been single dwellings. Though upon some of them charred wood was found, nothing else was discovered except a mortised piece of timber, which might have been drifted there; and in one, inserted under the upper layer of brushwood, a large oak beam, measuring 8 feet long by 3 in circumference.

"This group of small islets was close to the shore. They had, however, been surrounded by water at the time the level of the loch reached the highest beach-mark. I could not discover any causeway or piled connection with the shore.

"Near the north margin of the loch, a canoe was found in the mud. It measured 25 feet in length, and was strengthened by a projecting cross band towards the centre, left in the solid in hollowing out the inside."

RELICS FOUND AT DOWALTON.

The relics found in the course of these investigations at Dowalton Loch were presented to the Society of Antiquaries of Scotland by Sir William Maxwell of Monreith in 1865, and they are now deposited in the National Museum, Edin-

burgh.   The following description of them is taken from the
Proceedings of the Society, vol. vi. p. 109:—

Square-shaped stone, 5 inches in length, 1 inch in breadth,
and $\frac{5}{8}$ inch in thickness, and tapering to a point $\frac{1}{8}$ inch
square; probably a whetstone.

Three bronze basins: one measures 10 inches in diameter,
and 4 inches in depth.   It is formed of sheet metal, fastened
by rivets, with portions of an iron handle.   This pot or basin
shows several patches or mendings (Fig. 11).

Fig. 11. — Bronze Basin (height 4 inches).

Another vessel of bronze measures 12 inches in diameter, and 4 inches in depth.   It appears to have been made
by hammering it into shape out of one piece of metal.

The third vessel measures 12 inches in diameter, and 3
inches in depth, and is also formed out of one piece of metal.

Fig. 12 — Bronze Basin (height 3 inches).

On its upper edge is a turned-over or projecting rim, 1 inch
in breadth (Fig. 12).

Pot or patella of yellowish-coloured bronze, with a handle
springing from the upper edge, 7 inches in length, on which
is stamped the letters ·CIPIPOLIEI.   At the further extremity is a circular opening.   The bottom is ornamented
by five projecting rings, and measures in diameter 6 inches ;
it is 8 inches in diameter across the mouth; the inside
appears to be coated with tin, and has a series of incised
lines at various distances.   The vessel is ornamented on the

Fig. 13.— Bronze Pot (height 5½ inches).

outside opposite to the handle by a human face in relief,
surrounded by a moveable ring, which could be used in
lifting the pot (Fig. 13).

Bronze ring, measuring 3½ inches in diameter, which
passes through a loop fastened to a portion of broken bronze,
apparently part of the upper edge of a large bronze vessel,
the ring having formed one of the handles (Fig. 14).

Small very rude clay cup or crucible, 2 inches in height
(Fig. 15).

Bronze implement, being a short tube 1 inch in length,
with a projecting rim at one extremity, which is 2 inches in
diameter.   It is not unlike in shape to the socket portion
of a modern candlestick.

Bronze penannular ring or brooch, 1¾ inch in diameter, with bulbous extremities (Fig. 16).

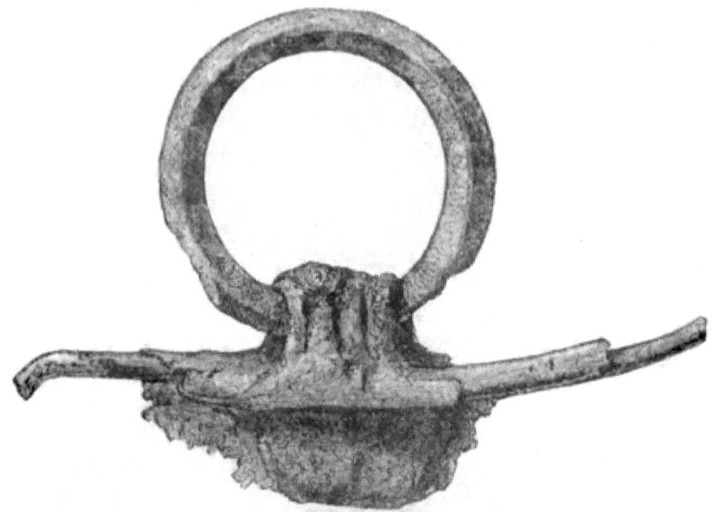

Fig. 14.—Bronze Ring (⅓).

Small plain bronze ring, 1 inch in diameter.

Small portion of bronze, probably portion of a vessel.

Fig. 15.—Crucible (½).

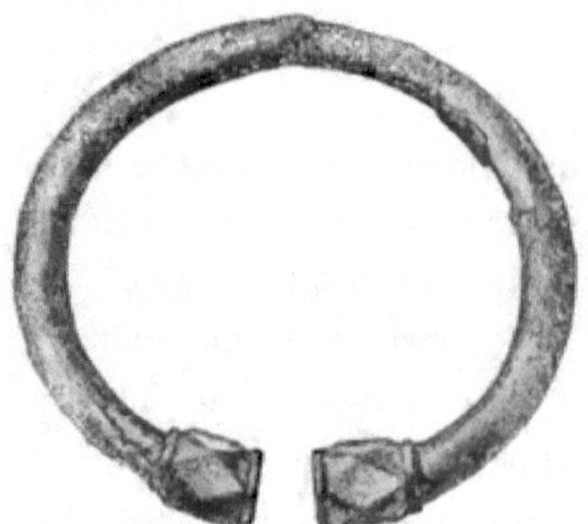

Fig. 16.—Bronze Penannular Brooch (½).

Small bronze plate or ornament, 1 inch in length, having a projecting tongue at three of its corners, each projecting portion being pierced with a hole through in its centre.

Two iron axe-heads: one with a square-shaped head,

which tapers to a sharp cutting face, and measures 6½ inches long; it has a large perforation close to the square head for receiving the handle (Fig. 17). The other measures 6 inches

Fig. 17.—Iron Axe (½).   Fig. 18.—Iron Axe (½).   Fig. 19.—Iron Hammer (½).

in length. The perforation for the handle is near the centre; and one end has a sharp cutting face, the other a blunt rounded extremity, or head (Fig. 18).

Iron hammer-head, 8½ inches in length, with hole in the centre for handle; the head is square, and tapers slightly to a blunt face (Fig. 19).

Several masses of iron slag.

Wooden boat paddle, the blade measures 2 feet 4 inches in length, by 10 inches in breadth, and 1 inch in thickness. It has a short rounded handle, measuring 7 inches in length.

Half of a ring, 3 inches in diameter, formed of white glass or vitreous paste, and streaked with blue (Fig. 20).

Half of a similar ring, formed of yellow-coloured glass or vitreous paste.

Large bead, measuring 1½ inch in diameter. The centre portion is formed of blue glass, of a ribbed pattern. The central perforation or opening is formed of a tube of bronze, and the edge of both sides of the perforation is ornamented by three minute bands of twisted yellow glass (Fig. 21).

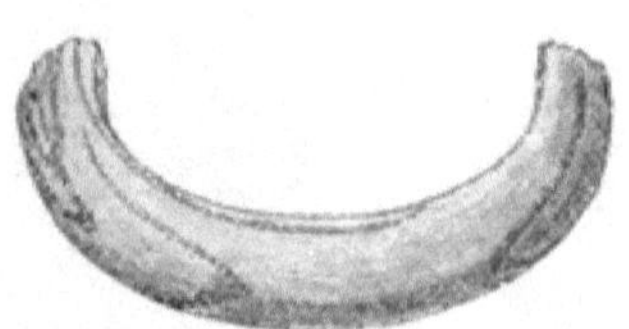

Fig. 20.—Portion of Ring of Glass (½).

Fig. 21. Bead
(length 1 inch, height 1¼ inch).

Bead of earthenware, ¾ inch in diameter, of a ribbed pattern, and showing traces of green glaze (Fig. 22).

Small bead of vitreous paste, of a white colour with red spots, and measuring ¼ inch in diameter (Fig. 23).

Fig. 23.—Bead. Fig. 22.—Bead. Fig. 24.—Bead.
(All actual size.)

Amber bead, ¾ inch in diameter.

Half of a small bead, measuring ¾ of an inch in diameter, of white glass streaked with blue (Fig. 24).

Small portion of blue glass.

Portion of a leather shoe, measuring 7 inches in length, and 3½ inches in its greatest breadth, nearly covered with ornamental stamped patterns (Fig. 25).

Fig. 25.—Portion of Shoe (length 7 inches).

Besides the above list there were found five canoes, five quern-stones, and several whetstones.

On the 14th of March 1881, R. Vans Agnew, Esq. of Barnbarroch, presented to the Museum of the Society of Antiquaries of Scotland, a brooch or ornamental mounting of bronze, found in Dowalton Loch, Wigtownshire, of which Fig. 26 is a representation. It is ornamented with trumpet-shaped spaces, probably filled with enamel, and measures 2 inches in diameter. Mr. Vans Agnew gives the following account of the circumstances in which it was discovered:—

Fig. 26.—Bronze Ornament (2 inches in diameter).

" The bronze ornament or brooch was found last summer in the bed of the Loch of Dowalton by Master Alexander Gibson, grandson of Mr. Alexander Cumming, the venerable tenant of the farm of Stonehouse, on the shore of the lake. It was then seventeen years since the lake was drained. I have not been able to ascertain the exact spot where it was found, but it was not far from the site of some of the crannogs."[1]

REPORT ON OSSEOUS REMAINS.

The following is Professor Owen's report of the bones which were submitted to him for examination:—

"The bones and teeth from the lake-dwellings, submitted to my examination by Lord Lovaine, included parts of the ox, hog, and goat. The ox was of the size of the *Bos longi-frons*, or Highland kyloe, and was represented by teeth, portions of the lower jaw, and some bones of the limbs and trunk. The remains of the *sus* were a lower jaw of a sow, of the size of the wild boar, and detached teeth. With the remains of the small ruminant, of the size of the sheep, was a portion of a cranium with the base of a horn core, more resembling in shape that of the he-goat. Not any of these remains had lost their animal matter.—R. O."

LOCH KIELZIEBAR, ARGYLLSHIRE.

In December 1867 a paper, by the Rev. R. J. Mapleton, Corr. Mem. S. Ant. Scot., Kilmartin, was read at the meeting of the S. A. Scot., describing an artificial island in Loch Kielziebar, near the Crinan Canal. The author thus sums up his observations:—" Altogether, I think that it is evident that the crannog was entirely composed of rock and walling, with the middle part filled up with smaller stones: that there existed considerable works of wood on the east, south, and west sides, at least, but whether a rampart outside, or a building on the structure itself, is not quite clear; that there

---

[1] *Proceedings Soc. Antiq. Scot.* vol. iii., new series, p. 155.

was a partial causeway, now under water, and the interval either filled in with brushwood, or passed over in a canoe."—(*Proc. Soc. Antiq. Scot.* vol. vii. p. 322.)

In June of the following year Mr. Mapleton gives a description of stockaded remains discovered twelve years previously upon the partial drainage of a fresh-water loch at Arisaig, in the parish of Ardnamurchan, Inverness-shire. This loch was of an irregular oval form, and lay in a comparatively level tract of land, with very low braes at a short distance from its shores. It communicated with the sea by a small burn. The crannog was of a rectangular shape (43 by 41 feet), but owing to the surrounding mud it was impossible to ascertain how the foundation of the crannog had been formed. "Outside of the building is a range of sharpened posts, fixed in the bottom of the loch, and inclining inwards towards the crannog, leaving a space of about 3 feet of water between them and the building. These posts are beautifully pointed, being quite round towards the ends, as though made by small sharp instruments. We counted eight still standing on one side. The crannog appears to have been formed altogether of very large round logs, or rather of trees with the bark left on, and the side branches neatly cut off. They are of various lengths: one that we were able to measure being 29 feet long, and 5 feet in circumference, at about 2 feet above the base. Another log was closely fitted to this, so as to extend through the whole breadth of the building. The ends did not overlap, but had been neatly cut or worn off, so as to be placed quite close to each other.

"We tried to dig down into the structure, and found at least four layers of these large trunks placed very regularly across each other. We could not dig deeper, as the water began to ooze in; but by using a probe, we felt timbers at a depth of 8 feet below our digging. The wood is chiefly oak, but there are some logs of birch. . . . On the surface were several large flagstones, especially in three spots. These bore strong marks of fire, and the logs on which they rested were much charred. Beneath and around them we found charcoal, several small pieces of calcined bone, shells of hazel-nuts, and one very small chip of flint, together with several rough angular pieces of white quartz. At each of the four corners of the structure there were two sharpened stakes inclining towards each other and the building, leaving a small space between them; and at one end (viz. the south-east) there was one large log of oak 39 feet long, and 5 feet 6 inches in circumference at the base. Two great logs were nicely rounded off at the end, and a hollow was scooped out in the wood, about 2 or 3 inches deep, and 4 inches broad.

"Upon rowing up to the structure, when it first appeared above the surface of the falling water, the men first came to a kind of rampart, that ran on all the four sides, about 3 feet distant from the structure, and about 18 inches higher than the apparent level of the floor of the crannog. This was formed by large trees that were kept in their place by the upright sharpened posts, whose sharp points projected about 1 foot above the trees. The ends of these trees were scooped out in the same manner as the two that still remain; and they were firmly fixed in their places between the two sharpened posts at each corner, which fitted into the hollow made by the scooping. No signs of a causeway were observed, neither could we detect any symptom of one, though we carefully probed the mud all round. 'Lord Abinger informed me that when a loch on his property, Torlundie, Fort-William, was drained, there was a kind of structure with timbers in it, which were unfortunately scattered and destroyed, as Mr. Stuart had not then made known the existence of crannogs in Scotland, and drawn attention to them.'"—(*Proceedings*, vol. vii. p. 516.)

ARTIFICIAL ISLANDS IN MULL.

In June 1870, the following note by Farquhard Campbell, Esq. of Aros, Mull, was read at the meeting of the Society of Antiquaries of Scotland.—(*Proceedings*, vol. viii. p. 465.)

" The loch called *Na Mial*, in English *Of Deer*, is about a mile south of Tobermory, and about 150 feet above the level of the sea, and 50 acres in extent. There was in the loch one of the artificial islands which are found in almost all the lochs of Mull. I drained the loch, which was only about 6 feet deep of water, blasting a passage through whinstone rock 20 feet deep. The mud under the water is of great depth. Of course, we had to make deep drains round the loch to catch the water. On coming with the drain to the edge of the loch, opposite this island, a large canoe was found 4 feet under the surface of the mud. The canoe was of black oak, 17 feet in length and $3\frac{1}{2}$ feet beam, quite fresh and sound. Several canoes of a smaller size were also found, but near the surface of the mud, and in a half-decayed state. Three boats of modern *clinker-built* construction, of whose history none of the natives had any knowledge, were also found. I had the large canoe dug out of the mud and put into the sea, in order that, being saturated with salt water, it might be preserved from cracking. There is another loch on my property which has two of these artificial islands. The loch is large—about 1500 acres. I may also mention that, close to the site of the large canoe, I found *a stone causeway laid upon oak-trees*. This was at the same depth under the surface of the mud (viz., about 4 feet). This causeway led direct to the artificial island, which was formed of a quantity of loose stones, on the only rock near the surface of the water in the whole loch."

LAKE-DWELLINGS OF LEDAIG AND LOCHNELL, ARGYLLSHIRE.

Dr. Angus Smith, F.R.S., in a communication to the

Society of Antiquaries of Scotland in 1871, describes, among other antiquities near Loch Etive, lake-dwellings at Ledaig and Lochnell, the former of which, notwithstanding the limited and inadequate inspection it has undergone, presents some features of interest, which the reader will find in the following extracts from Dr. Smith's report.—(*Proc. Soc. Antiq. Scot.* vol. ix. pp. 93 and 105.)

"About one hundred and twenty years ago a company from England, engaged in working iron, had diverted a stream from this to the east, and made dry ground where was a lake.

"The space that called forth interest was scarcely distinguishable from the rest of the moss. A little attention, however, showed a depression. The whole was of a brownish-green colour, but in the middle of the depression, where had been the old lake, there was a part greener than the rest. It was of an oval form, about 50 feet long, and 28 feet broad. The outer part had a double row of tufts, as if two walls had existed. I expected piles at these places, but the whole was soft and consisted of turf only. On digging down, about $3\frac{1}{2}$ feet, we came to wood, consisting of young trees from 6 to 8 inches in diameter, lying packed closely together. Under these there was another larger layer crossing, and under these again more. There seemed four all along the building. This was opened in three parts, and the same layers of wood were seen. . . . .

"At the east end of the oval was an elongation not surrounded by the turf mound. I believe the foundation extends along it. I suppose this to have been a platform before the door, a place for the inhabitants to sun themselves, and a landing and disembarking spot. (This platform was afterwards found to extend all round.)

"In the middle nearly, but a little to the westerly end, of the oval house was the fireplace. It is higher than the rest of the space. It was here that the bones were found, with shells and nuts. Under a few inches of a white powder is the hearth. It consists of four flattish stones; under the stones are to be found more peat ashes and some few remnants, but very few, of the substances connected with food. Under the ash was a floor of clay about 6 inches thick."

Dr. Smith, having resumed excavations here on a subsequent occasion, remarks (*Proc. Soc. Antiq. Scot.* vol. x. p. 82): " A little more was exposed this year, and a third fireplace found at the north-western end. On each side, a little towards the front, was a raised seat. This was a bank of earth on which were placed flattish stones. These were the arm-chairs of the inhabitants. Amongst the rubbish outside the wall were found two or three piles, the meaning of which is not yet made out. Two broken combs made of wood were obtained, one of which is shown in the annexed woodcut (Fig. 27).

Fig. 27. —Wooden Comb (⅔).

" A piece of wood with a cross burnt on it caused a good deal of interest. This kind of cross is not uncommon in the older Irish forms. It is a Greek cross with crosslets, and has been imagined to indicate a time before the Latin Church entered."

A small island in Lochnell is supposed by Dr. Smith, after a slight examination, to be another lake-dwelling.

CRANNOGS IN WIGTOWNSHIRE.

In the same year (1871), the Rev. George Wilson, Glenluce, contributed a paper to the Society of Antiquaries of Scotland (*Proceedings*, vol. ix. p. 368), on the Crannogs and Lake-Dwellings in Wigtownshire, from which it would appear that all the lakes in this locality were once literally studded with these island habitations. He enumerates no less than ten lakes, each of which contained one or more crannogs. The abundant remains of stakes, mortised beams, and the occasional discovery of a "paved ford" connecting the islands with the shore, sufficiently indicate their structural formation; but beyond this, and the important fact of their existence in such numbers in the district, they present nothing of a novel or special character calling for a more detailed notice here. (See tabular statement, p. 245.)

The relics from the Wigtownshire crannogs, besides those

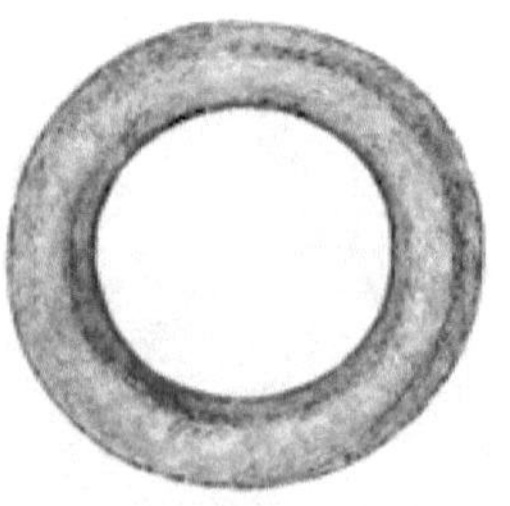

Fig. 28.—Stone Ring (¼).

already noticed from Dowalton, are not many. They are two granite querns found near a stone causeway leading to the crannog, a stone ring, ¾ inch internal diameter (Fig. 28), and a spindle-whorl of clay slate, 2 inches in diameter, from a crannog in Barlockhart Loch.—(*Proc. Soc. Antiq. Scot.* vol. iii., new series, p. 267, and vol. xi. p. 583.)

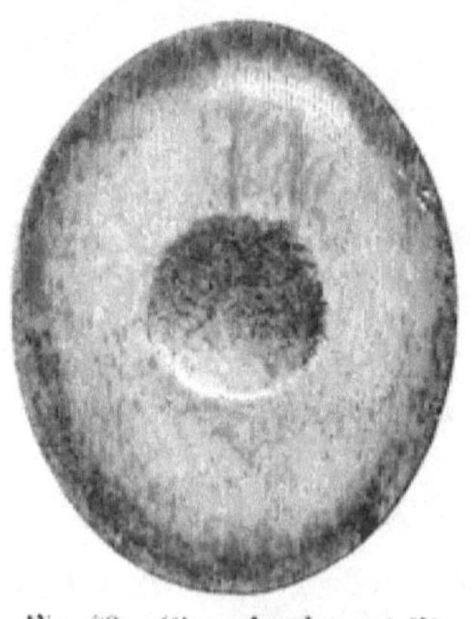

Fig. 29.—Stone Implement (½).

Regarding stone implements, with circular central hollows wrought on each face, one of which (Fig. 29) was found on a crannog in Machermore Loch, Mr. Wilson writes thus:[1]— "These are of two types, elongated and oval, approaching a circular form, and I wish to direct attention to them, because, as yet, only eight have been reported in Scotland, seven of them being from Wigtownshire."

[1] *Proceedings Soc. Antiq. Scot.* vol. ii., new series, pp. 127, 128.

On a later occasion, June 15th, 1881, Mr. Wilson, writing on the same subject, says :—" In the volume of the Proceedings for 1879-80, at pages 127-129, I have described seven of these stones, and have stated that only one specimen has been reported from any other part of Scotland.   I now direct attention to eleven more from Glenluce and Stony Kirk added to the Museum, making eighteen from Wigtownshire." [1]   (See notice of another, found on the crannog in Lochspouts, at page 173.)

One of the crannogs referred to by Mr. Wilson, viz., that in " Loch Inch-Cryndil," or Black Loch, was about the same time subjected to a careful examination, a report of which was drawn up by Charles E. Dalrymple, Esq., F.S.A. Scot. (*Proc. Soc. Antiq. Scot.* vol. ix. p. 388), from which I quote the following illustrative extracts :—

" The island is oval in shape, 180 feet long, and 135 feet broad in the widest part.   It has tolerably deep water round it, excepting towards the nearest shore, a distance of about one hundred yards, where, in dry seasons, it does not exceed 6 or 7 feet. . . .

" In the middle of the island, which is thickly covered with trees of thirty or forty years' growth, but with a few much older toward the south end, a circular mound appeared, resembling a low tumulus, 45 feet in diameter, rising in the centre to about $3\frac{1}{2}$ feet in height, round the edges of which there were, in some parts, traces of a low wall of three or four courses of small stones, like a miniature dike.   The island rises gradually from the water to the base of the mound, which at that season (the beginning of October) was about 18 inches above it, so that the top of the mound, which was the highest part of the island, was then about 5 feet above the loch.   Spacious cuttings were made in the centre, afterwards extended to the edge of the mound in

<hr>

[1] *Proceedings Soc. Antiq. Scot.* vol. iii., new series, p. 266.

various directions, with the following results :—The island proved to have been a crannog, formed apparently on a shoal in the lake, composed of shingle over blue clay, the object having obviously been to raise a platform which would be above the water even when the lake was at its fullest, as, even at the present time, there is a considerable rise in the wet months, although pains are taken to keep clear the outfall from the loch. The mound was found to be of earth and stones mixed, extending beneath which, at a depth of 5 feet in the centre, but decreasing in depth towards the edge, was found a flooring of trunks of trees, oak and alder, in two layers, crossing each other at right angles in some places, in others lying rather confusedly. These were mostly not more than 6 or 8 inches in diameter, but one solitary trunk of an oak, near the centre, lying at a higher level, and possibly the remains of a hut or other superstructure, was fully two feet in diameter, although much decayed. These layers of wood were traced as having covered a circular space about fifty feet in diameter, thus agreeing nearly with the size as well as the shape of the mound. . . . The extent of the mound would appear to have been that of the crannog proper, but the existence of a solitary oak pile, 50 feet from it, on the weather side of the island, makes it probable that either a breakwater had been placed there, as was also supposed to be the case in Dowalton Loch, or a ' chevaux-de-frise ' of sharp-pointed stakes for defence.

" At different levels, from that of a few inches above the timber flooring to 3 feet higher, and over the whole mound, were found many fireplaces, one or two covered over with two long stones, leaning against each other lengthways, like the roof of a house, but most of them formed by placing two long narrow stones (fragments of the rock of the district, which breaks off easily in that form) parallel with each other, leaving a space between, which was paved with small stones and formed a hearth. Large quantities of bones of animals, mostly more or less burnt, and, whether flat or round bones, frequently split, were found mixed with the ashes and charcoal which lay in and around these hearths, in some places extending over wide spaces, which were marked, also, by masses of burnt yellow clay."

At different levels, in different parts of the mound, were found the following objects,[1] the description of which I take from *Proc. Soc. Antiq. Scot.* vol. ix. p. 381 :—

Double-margined comb of bone, imperfect, $2\frac{7}{8}$ inches across, formed of separate pieces, enclosed between two transverse slips of bone, fastened with three iron rivets, and

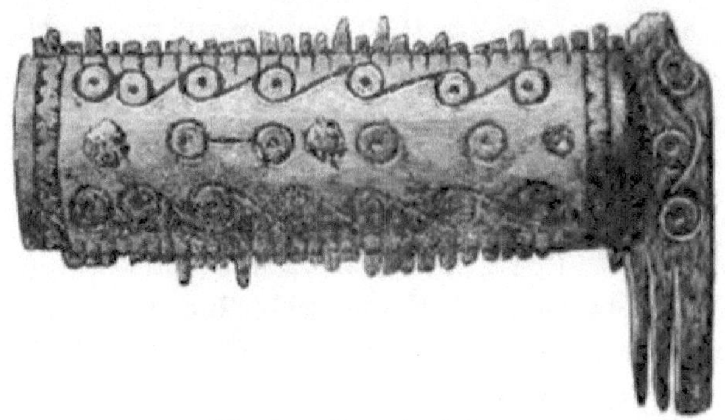

Fig. 30.—Bone Comb (½).

ornamented with a central row of dots and circles, and two similar rows at the side of the cross piece, having a running scroll pattern connecting them. A zigzag ornament forms a band across the end (Fig. 30).

A flat loop of bronze, $1\frac{1}{4}$ inch in diameter.

Part of the rim of a large vessel of cast bronze, 3 inches in length.

Portion of an armlet, of greenish glass, with a blue and white twisted cable ornament running round it.

---

[1] These relics were sent as donations to the Museum of the Society of Antiquaries of Scotland, by the Right Hon. the Earl of Stair, in 1872.

Copper coin much defaced.

Copper bodle of Charles II.[1]

CRANNOG AT TOLSTA, LEWIS (1874).

The following account of a crannog is from a letter by
Peter Liddle, Esq., to the secretary of the Society of Antiquaries of Scotland (*Proceedings*, vol. x. p. 741) :—

" In a lake recently drained at Tolsta, I have examined
a crannog which seems to me to possess some interest. A
drain has been cut through part of the crannog, which
affords a section of its construction. At the outside there
is a row of piles 5 or 6 inches diameter, then large stones,
then another row of piles, then heather and moss—the whole
covered with earth and gravel. The remains of three houses
built of unhewn stones are still visible upon it. All round
the crannog, but inside the outer row of piles, there is an
immense quantity of shells, plentifully intermixed with
bones, ashes, and twigs of trees. The shells are those of the
ordinary edible shell-fish, the mussel being the most common.
The bones are chiefly those of deer, and the small Highland
sheep still found on the island. The only implement I
found was part of a stag's horn, with the brow-antler
thinned. Three hollowed stone vessels or knocking-stones
were found on the surface, but they were destroyed or lost
sight of. A causeway of large stones under water led to the
crannog."

---

[1] The existence of these coins in the crannog is thus explained by
Mr. Dalrymple :—" It is known that the island has been planted two or
three times, and that considerable quantities of soil and stones have been
added to it. The two feet of soil which covered the uppermost remains,
and which so much raised the centre of the crannog, was probably added,
in great part, about 1720, when Field-Marshal the Earl of Stair laid out
the grounds of Castle Kennedy. Some of these operations may, to some
extent, have disturbed the remains. They would, at all events, account
for the modern coins found so far below the surface."

Loch Lotus, Kirkcudbrightshire.

During the summer of 1874, a canoe (Fig. 31) was discovered in Loch Arthur, or Lotus Loch, in the stewartry of Kirkcudbright, in the vicinity of a small artificial island which is thus described by Rev. James Gillespie (*Proc. Soc. Antiq. Scot.* vol. xi. p. 21): "When fully exposed to view by the trench which was dug around it, the canoe was seen to be of great size, ornately finished, and in a fair state of preservation. It had been hollowed out of the trunk of an oak, which must have been a patriarch of the forest, the extreme length of the canoe being 45 feet, and the breadth

Fig 31.—Canoe found in Loch Arthur.

at the stern 5 feet. The boat gradually tapers from the stern to the prow, which ends in a remarkable prolongation resembling the outstretched neck and head of an animal. When excavated this portion of the canoe was entire. At the neck of the figure-head, there is a circular hole about 5 inches in diameter from side to side. At the prow a small flight of steps has been carved in the solid oak from the top to the bottom of the canoe. The stern is square, and formed of a separate piece of wood, inserted in a groove about an inch and a half from the extremity of the canoe. The stern-board board when found was in a fragmentary condition, so that it is impossible to say whether it consisted of one or several planks.

"Along the starboard side (which when found was in good preservation except near the stern), there could be traced seven holes about 3 inches in diameter. The three front holes were nearly perfect, but at the stern the side was so broken that only the lower parts of the holes could be observed. They are about 5 feet apart, and the front hole is about that distance from the prow—the last being about 7 feet from the stern. There are three holes pierced through the bottom at irregular intervals.

"In connection with the discovery of this canoe, it is worthy of remark, that on the opposite side of the lake, between three and four hundred yards from the spot where the canoe was found, there is a small circular island which is evidently artificial. It is about 100 feet in diameter, and is approached by a stone causeway about 30 yards long, which was laid bare last summer by the lowness of the lake. The artificial nature of the island may be seen by the remains of the oaken piles driven in in rows, with horizontal beams between, which can still be traced in the water round the north-east and south sides. The lines of two small enclosures can be followed on the south side of the island.

"No excavations have yet been made on the island, but ashes and other signs of fire were found many years ago."

CRANNOG IN THE LOCH OF KILBIRNIE, AYRSHIRE.

At the meeting of the Society of Antiquaries of Scotland in June 1875, Robert Love, Esq., F.S.A. Scot., gave a description of a crannog in the loch of Kilbirnie, of which the following is a condensed account.—(*Proceedings*, vol. xi. p. 284.)

"There was a little island in the upper end, and near the north-west corner of this loch; and most who knew it when entire, 50 or 60 years ago, are agreed that it was essentially circular, although some little pointed towards the south. It was elevated, at least in modern times, above the water of the loch in its ordinary state, from 2 to 4 feet; and on the surface was entirely overlaid with stones of the boulder sort, not large, and which might have been got on the margin of the lake. Some say that beams or logs, and piles of wood were noticed during protracted droughts on or along the margin of the island, but if they were, it notwithstanding never occurred to any one that the island was other than natural. In the summer of 1868, however, its artificial nature became quite evident. This was occasioned in consequence of the slag from the furnaces having been for several years, and in great bulk, deposited within the loch to the west of and behind this island, which sunk down through the soft yielding mud deposit there, which is of the great depth of 30 or 40 feet, a fact that was ascertained by borings near the site of the furnaces. This had the effect, while it overlay and bore down that part of the island which is towards the west, of moving the east portion of it forward and into the loch, and, at the same time, of upheaving it so that it was elevated considerably above the water. In consequence, this part spread hither and thither and split up ; many fissures were the result, both in the artificial deposits and in the underlying mud, which were of a depth that varied from 4 to 6 feet ; and it was by means of these that the various artificial strata became disclosed.

" It has been said that the surface of the island throughout was overlaid or paved with stones. The depth of these was not great, possibly not more than from 1 to 2 feet, there not being in any part that became visible more than two courses. Wood ashes were discovered on the surface—a portion being also found a little below, and some of the stones at one part, in particular the fragments of a sandstone flag, bore distinct evidence of the action of fire ; and it was supposed that this flag might have been the hearth of some structure reared on the surface. These stones are to be held as the uppermost artificial stratum. The next in descent was a layer of large coarse water-borne gravel mixed with finer sand, which was of the depth of from 18 inches to 2 feet. The third layer was brushwood, boughs of trees, among which the hazel predominated, ferns, etc. etc., but the whole was so compressed as not to manifest a greater depth than about 6 inches. The fourth layer was beams or logs of wood, some of which were nearly 2 feet in diameter, although the greater number was less. These seemed laid down horizontally, and so as to cross or intersect each other, similar to a raft of wood; some of them showed that they had been mortised or checked into each other, or into vertical piles, and that the tenons when inserted had been fastened by wooden pins, and in one or two instances by large iron nails.

" The whole of this wood-work, however, when exposed, was in a greatly disturbed and loosened condition from the movement and upheaval of the structure; and, in consequence, what space in depth these cross-beams occupied was not ascertainable. Then the fifth and lowest stratum was the underlying mud, which was fine, pure, and free of stones, and not at all like boulder clay. Besides, there was manifested as having been planted on the surface, one if not more wooden structures, houses or huts they might be, small in size, and one of which at least was in the form of a parallelogram, having been constructed of small round posts of wood used in forming the sides and ends. How it had been roofed did not appear. There were seen also bits of bone, as those of birds, as well as a few teeth, similar to those of the cow or ox. Trees, for the most part of a low stature, were over all parts, as well as reeds and other coarse grasses which sprang up between the stones on the surface.

" Then as regards the *margin* of this island, it appeared to have been palisaded; at least this was the case on its northeast side—that which only was visible. The piles used for this purpose were apparently of oak, and not great in girth; they were driven down into the mud bank as the foundation; and on these, as well as upon the beams, the cutting of an edged tool, not a saw, was quite distinct. Within these vertically placed piles, and resting on the surface, stones, it is said, were placed, which was the case more certainly around the whole margin. It is also said that stones were even placed outside of these piles, in a row, and on the very margin; but it is only probable that *outwith* this row there had been an outer course of piles, by means of which the stones were kept in position, but which, from weathering, had gone into complete decay.

" It is known that this island was approachable by means of a kind of stone causeway which led from the north-west margin of the lake. According to the report of those who saw it often, it was only of the breadth of 2 or 3 feet, and was never visible above the water of the loch, which on either side is said to have been 6 or 7 feet in depth. It is not said that this causeway was protected or fortified in any way by piling. It was near the south end of this causeway, along the north-east margin of the island, that in 1868 several canoes or boats, as many it was believed as four, in a less or more entire condition, were discovered. Only one of these, however, when found, was partly entire, and it even wanted some 2 feet at the bow to render it complete. But as this canoe, formed out of a single tree, and the bronze utensils which were found imbedded in mud within it, have been well described in Mr. Cochran-Patrick's paper, printed in the Society's Proceedings (vol. ix. 385), none of these need now be referred to, further than to say that the pot, the repair or clouting of which was with *iron*, is not by any means uncommon in shape."

The following is an extract from Mr. Cochran-Patrick's description of these relics, above referred to :—

" The canoe was discovered lying about 20 feet north of a small artificial island—itself an object of great interest, but now unfortunately overwhelmed by the advance of the iron-stone rubbish at the south-western end of the loch. It was hollowed out of a single tree, and was about 18 feet in length, 3 feet in breadth, and close on 2 feet in depth. It was broadest at the stern, which was square, and tapered towards the bow, and was entire, with the exception of about 2 feet broken off the narrowest end. There were indications that a hole in the bottom had been mended, and

Fig. 32.—Lion Ewer, the property of W. J. Armstrong, Esq., found in a canoe in the bottom of the Loch of Killbirnie (8½ inches high).

some wooden pins were in it which may have been used for this purpose, or for fixing at the side what is described to me as a sort of bracket. In the mud which filled the hollow of the canoe were found a lion-shaped ewer (Fig. 32) and a three-legged pot, both made of bronze, and also a thin plate or piece of metal which cannot now be recovered.

" The ' lion ' stands 8½ inches from the ground at the highest part, is 8 inches in length and 8½ in girth round the body, and weighs 4 lbs. It is made of a yellowish bronze, and seems to have been used for holding liquid. It bears a striking resemblance, though smaller and less ornamented, to one figured and described at p. 556 of Wilson's *Prehistoric Annals of Scotland* (edition 1851). It will be observed that the one now shown wants the curious ornament projecting from the breast, though the place where it has been inserted is quite apparent. The bronze pot is 11 inches across the mouth, stands 14 inches high, and weighs 28 lbs. It resembles what are often called Roman camp-kettles. There are indications of its having been ingeniously mended."

General Remarks.

This concludes a brief historical and descriptive sketch of ancient lake-dwellings, as known in Scotland previous to the excavation of the Lochlee Crannog in 1878-9, a full report of which will be found in the next chapter. From this sketch it will be seen that, during the interval between the publication of Dr. Stuart's paper in 1866 and the above date, if we except the occasional notice of the discovery of a new site, comparatively little has been done by way of furthering the systematic exploration of their widely-scattered remains. With the formation, however, of the Ayrshire and Wigtownshire Archæological Association, a new epoch in antiquarian research may be said to have dawned on the south-west of Scotland. One of the features of this Association is the prominence given to *practical explorations* as a means of investigating the prehistoric remains of the district, the beneficial result of which may be estimated by the fact, that, with a trifling exception, all the discoveries recorded and illustrated in the following pages are due to its inspiration, and have actually appeared, in the first instance, in its publications. These, however, constitute but a small part of the investigations conducted under the guidance and auspices of this most active Association.

CHAPTER III.

REPORT OF THE DISCOVERY AND EXAMINATION OF A CRANNOG
AT LOCHLEE, TARBOLTON, AYRSHIRE (1878-9).

Discovery of the Crannog.—The site of the Lochlee Crannog
was a small lake, now entirely dried up, which formerly
occupied portions of a few fields on the farm of Lochlee near
Tarbolton. The lake was surrounded by a gently undulating country, and lay in a hollow, scooped out of the glacial
drift, at an elevation of about 400 feet above the sea-level.
Taking a fair estimate of its former extent by a careful
examination of sedimentary deposits near its shore, it was
ascertained, from accurate measurements and levelling, that
its area was about 19 acres; but, owing probably to the
accumulation of moss and silt, it is known, in modern times,
to have been much greater, especially during winter.
Before it was artificially drained, some forty years ago, no
one appears to have surmised that a small island, which
became visible in the summer-time, and formed a safe habitation for gulls and other sea-birds during the breeding season,
was formerly the residence of man; nor am I aware of any
historical notices or traditions that such was the case; nor
does it appear to have attracted the attention of the poet
Burns, though he lived for four years on this farm in the
capacity of ploughman to his father, then tenant of the
place. The crannog was near the outlet of the lake, and the
nearest land, its southern bank, was about 75 yards distant.

When the first drainage of the place was carried out, the wrought wood-work exposed in the drains passing through the island, and especially the discovery of two canoes buried in the moss, attracted the attention of the workmen. The shop of a provision merchant at Tarbolton happened to be much frequented by the drainers, and in this way the shop-keeper, Mr. James Brown, came to hear of the finding of the canoes, and the conjectures of the men as to the artificial nature of the island. Mr. Brown, who seems to combine the true spirit of the antiquary with his business habits, never lost sight of the little island at Lochlee and the information he had ascertained regarding it, and on various occasions since, mentioned the subject to gentlemen who, he thought, were likely to take an interest in it. The recent re-drainage of the same locality revived Mr. Brown's curiosity about the structure of this island, now a slight mound in a field, and being himself unable, owing to the infirmities of age, to take any active part in inspecting it, he wrote a letter about the beginning of September to a gentleman at Ayr suggest-ing an inquiry into the matter; but as the latter did not seem inclined to take it up, a week afterwards he wrote a note to Mr. Anderson, of the National Museum of Antiquities in Edinburgh. This gentleman, recognising the importance of his information, immediately communicated with R. W. Cochran-Patrick, Esq. of Woodside, Hon. Secretary of the Ayr and Wigtown Archæological Association, who lost no time in visiting the locality, and at once discovered the true nature of the mound. Mr. Cochran-Patrick then sent a note to Mr. Turner, factor to the Duke of Portland, under whose supervision the drainage was being conducted, informing him of the discovery, and suggesting, in the interests of Archæo-logical Science, that an examination of the crannog should be made. Meantime these facts were communicated to me by Mr. J. H. Turner, and having had my attention already directed to Lake Dwellings in consequence of a recent opportunity I had of inspecting some of their relics preserved at Zürich, I also became interested in ascertaining the exact nature of the find at Lochlee. Next day Mr. J. H. Turner and I visited the locality, and in the course of a few more visits found ample evidence that the mound was really artificial, and had been at some former period the site of a human habitation. At the same time, as if to deepen our curiosity, a small canoe, hollowed out of a single trunk of oak, was dug up by the workmen out of the moss which formed the bottom of the lake. It was then kindly arranged by Mr. Turner, senior, that some excavations would be made so as to ascertain more accurately the structure of this

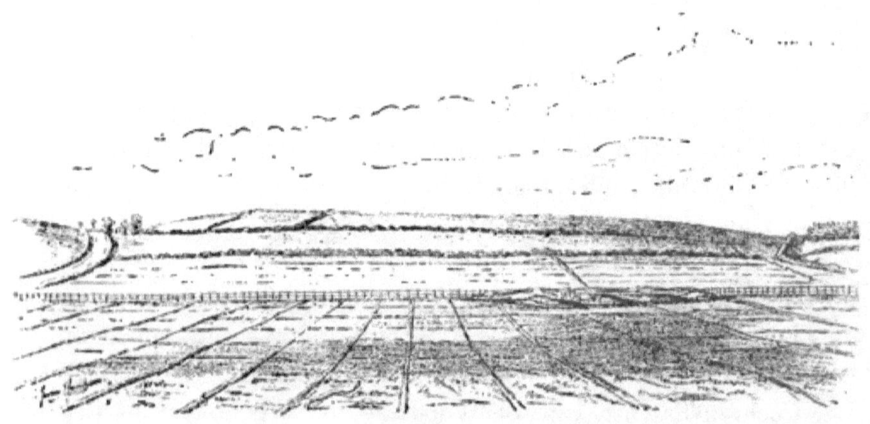

Fig. 33.--The Crannog after the Excavations were commenced.

mound. The general appearance which it presented after these excavations were commenced, as seen in Fig. 33, was that of a grassy knoll, drier, firmer, and slightly more elevated, than the surrounding field. Unfortunately, the large deep main drain which happened to pass through and cut off a segment of this mound, was filled up before attention was directed to its archæological importance, so that we lost the opportunity of inspecting the section which it presented to view. Upon careful inspection, however, we noticed towards the circumference of the mound the tops of a few wooden piles barely projecting above the grass, which at once suggested the idea that they might be portions of a circular stockaded island. Guided by these, I completed what we supposed to be the circumference of the original island, by inserting pins of wood where the piles were deficient. Following the line thus indicated, the workmen were ordered to dig a deep trench round the mound, but to leave whatever wood-work would be exposed as much as possible *in situ*. Accordingly, this trench was completed, and on the following day, 15th October 1878, systematic explorations were begun in presence of Messrs. Turner, J. H. Turner, Cochran-Patrick, Anderson, Dr. Macdonald (Ayr), and myself.

The Excavations.—The space enclosed by this trench was of a somewhat circular shape, and about 25 yards in diameter. The trench was from 5 to 6 feet deep, and in many parts quite studded with wooden piles, mostly upright, but some slanting. Some of those slanting outwards were forked at the upper end, as if intended to counteract outward pressure. At the bottom of the trench, particularly on the north side, were found various kinds of brushwood, chiefly hazel and birch, here and there trunks of trees, thick slabs of wood, and large stones. The most remarkable objects, however, were thick planks of oak about 6 feet long, with a large square-cut hole at each end. These were visible at various portions of the trench, and lying half-way down, some right across and others with one end sticking out from its inner side. At the north-east side there were two rows of these beams exposed, four in each row, and about 5 feet apart, measuring from the central line of each beam. One row was a little farther out than the other, and had upright piles, somewhat squarely cut, projecting through the holes.

These horizontal beams pointed towards the centre of the
crannog, and appeared to keep the upper ends of the upright piles in position (see Figs. 34, 35, and 36). Lying
underneath these beams, and at right angles to them, were
round logs of wood varying in length from 6 to 15 feet,

Fig. 34.—View of the Trench on the North side.

which being caught as it were by the upright piles, were prevented from falling outwards into the trench.   Conterminous
with the mortised beams, which were scarcely a foot under
the surface, there was a rude and much decayed platform of
rough planks and saplings resting on transverse beams of

split oak-trees. One of these transverse beams which I measured was 14½ feet long and 8 inches broad, and for a few inches at each extremity was not split, so that the portion thus left acted as a catch (for the planks above it), like the flange on the wheel of a railway wagon. Digging underneath this platform, we passed through a compact mass of clay, stones, beams of soft wood, and ultimately brushwood, underneath which, being on a level with the drain, we could not farther explore, owing to the oozing up of water.

Fig. 35.—Arrangement of Mortised Beams at north-east corner.[1]

We then commenced digging a few feet to the west of the centre of the mound, and soon cleared a trench from 3 to 4 feet deep, about a couple of yards broad, and directed almost due north and south. About 25 feet from the outer trench, measuring northwards, and 53 feet in the opposite direction, we came upon the south edge of a smooth pavement neatly constructed of flat stones. Judging from ashes,

[1] Before this sketch was taken some of the horizontal beams were removed.

charcoal, and small bits of burnt bones which were here observed, that this pavement was a fireplace, we thought it better in the meantime to leave it intact; so we formed another trench at a width of 8 to 10 feet, at right angles to the former, and just touching the southern edge of the pavement, which was continued eastwards till it touched the

Fig. 36.—Mortised Beam with portion of an upright (scale ½ inch to the foot).

platform already described. A circular trench was then made round this pavement, at a breadth of about 4 feet, leaving it, with its superincumbent soil, standing in the centre. We had thus a considerable space cleared out at a uniform level, with a small portion of the pavement visible, and an oval-shaped mass of soil about 4 feet in diameter

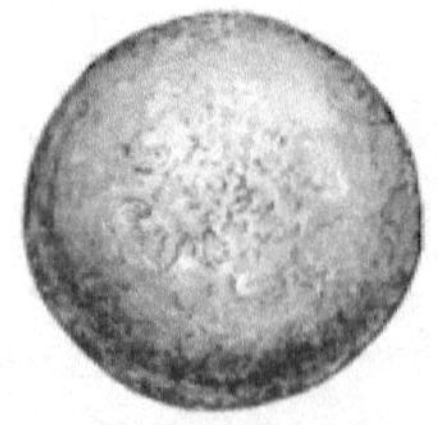

Fig. 37.—Quartz Pebble (½).

above the rest of it. In the course of these excavations we found three upper quern stones, portions of other two, a wooden vessel in two fragments, a large quartz pebble (Fig. 37), with markings as if made by a hammer on its surface, portion of a pointed horn (Fig. 91), some bones, one or two hammer-stones, and a boar's tusk.

Upon careful inspection we then discovered immediately above the pavement, at a height of $2\frac{1}{2}$ feet, and rather less than a foot from the surface of the mound, another pavement similar to the former. These pavements rested on layers of clay which extended several feet beyond them, and gradually thinned out towards the edge. On a level with the lower pavement we found the remains of a series of massive stakes with square-cut ends, which appeared to surround it. They were very much decayed, and it was difficult to ascertain their original number, but seven were noted, which

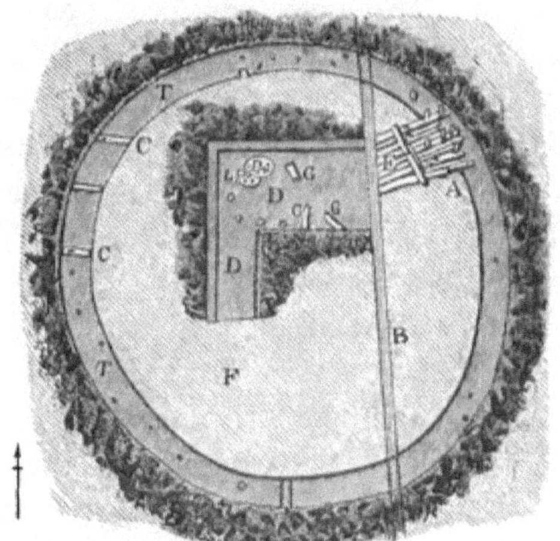

Fig. 38.—Diagram of Excavation.

T, Outer circular trench with stuff thrown outwards. DD, Trenches near centre of crannog. A, Mortised beams at north-east corner. E, Rude platform adjacent to mortised beams. P, LP, Upper and lower pavements or hearths, with stakes surrounding them. GG, Horizontal beams on level with lower pavement. B, Main drain passing through the mound. F, Undisturbed mound. CC, Two transverse beams lying across near the bottom of trench, with a square-cut hole in each, but not containing uprights.

were kept standing in position for some time. Two well-shaped plank-like beams were lying horizontally at the east side of the lower pavement, and on a level with it. The distance between these upright stakes varied from 2 to 4 feet, and, as already noticed, they were not pointed at their bases but cut across. One, indeed, we found to have a small portion projecting from the centre of its base, which neatly mortised into a hole formed by a piece of wood, a flat stone, and some clay. On a subsequent occasion, when digging lower, we came upon another of these stakes which had pressed down the portion of clay on which it rested nearly a foot. The lower pavement slanted a little to the south-west, and it was also observed that the bottoms of the stakes were somewhat lower in that direction. On the north side they came close to the pavement, but on the south extended about 5 feet beyond it. The upper pavement was about a foot nearer the outer trench, in the direction of the wooden platform already described at its north-east corner, and hence it only partially covered the lower. It was carefully built with stones and clay round a wooden stake, corresponding with the series of stakes on a level with the lower pavement, and the layer of clay underneath it extended eastwards over one of the horizontal beams above referred to. Both these pavements were neatly constructed of flat stones of various sizes, and about an inch and a half thick, and had a raised rim round them also formed of flat stones, but uniformly selected and set on edge. They were slightly oval in shape, and the major and minor axes of the lower one measured 5 and 4 feet respectively. Traces of other pavements between the upper and lower were observed, but before further examination was made the whole mass above the lower or first-discovered pavement was trodden down by visitors.

At this stage I have to record the loss of the active services of Mr. Cochran-Patrick, who hitherto took notes and sketches of each day's proceedings. In consequence of his absence, owing to a protracted illness, and the inability of the other gentlemen to attend, this duty now fell on my inexperienced shoulders; and in giving this short account of the work, I have only to say that, however imperfectly done, I have endeavoured, during very inclement weather, to procure as correct and faithful a record of the explorations as possible.

While making a tentative digging on the south side of the lower pavement, I ascertained that the soil underneath its corresponding layer of clay (which, by the way, extended much farther than any of the other layers) contained boars' tusks, broken bones, and charcoal. After digging for about 4 feet below the level of the pavement, we came upon a layer of chips of wood as if cut by a hatchet, and below this a thick layer of turf with the grassy side downwards. Water here oozed up, but with the spade I could readily distinguish that underneath the turf there were large logs of wood extending farther in all directions than I could then ascertain. With a pole we took the perpendicular height of the level of the surface of the upper hearth above these logs, and it measured exactly 7 feet 9 inches, so that the greatest depth of the accumulated rubbish since the logs were laid, i.e. about centre of mound, would be about $8\frac{1}{2}$ feet. I then determined to clear the soil entirely away round the fireplace down to these logs, still keeping the surrounding trench at the same breadth as before, viz., 4 to 5 feet. While this was being done we inspected the stuff as it was removed, though I now regret this was not done more carefully, and found a great variety of manufactured implements of various materials. Observe that the portion here referred to is well defined,—above by the layer of clay corresponding to the lower or first-discovered pavement, and below by the newly-discovered log pavement. It is fortunate that this was the case, as it turned out so prolific of relics that I have assigned to it the name of *relic-bed*. Amongst these were a spindle whorl (Fig. 66), two bone chisels (Figs. 69 and 70), and several pointed bone implements (Figs. 71 to 74), a polished stone celt (Fig. 55), a metal knife (Fig. 129), some implements of horn and wood, a fringe-like object manufactured of the stems of a moss (Fig. 151), and a great many hammer-stones. Close to the pavement, but about 2 feet lower, we extracted the skeleton of an animal like that of a goat or sheep, the skull of which was entire, and had short horn-cores attached to it. The relic-bed was made up of partially decomposed vegetable matters, and could be separated into thin layers; the common bracken, moss, parts of the stems of coarse grass, heather, and large quantities of the broken shells of hazel nuts, were frequently met with. The bones were generally broken as if for the extraction of their

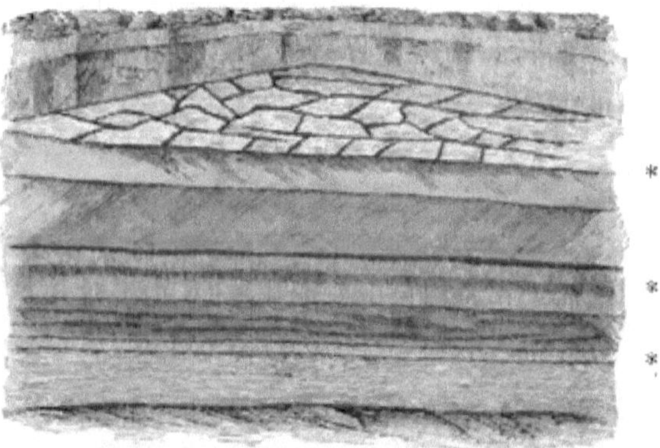

Fig. 39.—Perpendicular Section through the Hearths, showing structure of the first-discovered pavement. The asterisks indicate the position of the three lowest fireplaces, or stony pavements.

marrow. The bed of chips of wood was several inches thick, and extended more than half-way round, and had its maximum extent on the south-west side. The logs, all of which were oak, and cut at various lengths, from about 6 feet to 12 feet, seemed to radiate from the central line of the fireplace, like the spokes of a wheel. Underneath these logs were others lying transversely, and in some places a third layer could be detected by probing with a staff. None of these layers of logs were disturbed at this stage of the proceedings.

A perpendicular section made of the central mass left standing, just touching the southern edge of the first-discovered pavement, and looking towards the south, presented the appearance of stratified rocks of various colours, of which the above is a sketch. At the bottom is the log pavement; then in succession you see turf, clay, a black line of ashes; then again clay, another line of charcoal and ashes, and lastly, the pavement imbedded in a thick layer of clay. The upper pavement and intermediate section are not represented, as they were demolished by visitors some days previous to the taking of the sketch. Upon removing this central mass of clay and ashes intervening between the stony pavement and the log pavement, Dr. Macdonald and I made the important discovery that there were other two stony pavements corresponding exactly with the charcoal lines in the drawing. The one was 18 inches below the first-discovered pavement (or that figured in the drawing, and which has hitherto been called the lower pavement), and the other 16 inches still lower, and about a similar distance above the logs. Both these pavements were slightly oval in shape, about 4 feet in diameter, and beautifully built with flat stones and raised rims round them, precisely similar to the two already described. While in the act of demolishing these fireplaces we came upon another entire skull of a sheep or goat, with horn-cores attached to it, very like the one already mentioned, and found near the same place. At the north-east side, close to the fireplaces, were a few large stones built one above the other, and poised evenly with wedges of wood and stones. A little to the north of these stones, and about 4 feet from the base of the fireplaces, there was a portion of a large square-cut upright stake, a few feet long, resting on a flat circular board, like the bottom of a barrel, and supported by the log pavement. On the south side of the stones, and close to them, was a round flat piece of oak, with a hole in its centre, somewhat like a quern stone. My fist could just go through this hole, and when found it had a small plug of wood loosely fitting it. Near the same place portions of a large shallow dish made of soft wood, and a small bit of a three-plied rope of withs, were picked up. About 5 feet to the south of the centre of the pavements there was a portion of another upright stake resting on the log pavement. Although various other portions of decayed stakes and pins of oak were found while excavating within a few feet of the fireplaces, they were not so systematically arranged as to suggest the idea that they formed the remains of a surrounding hut, as was undoubtedly the case with those corresponding to the first-discovered pavement, and already described.

Before proceeding further, let me pause for a moment and endeavour to recall, in a few words, the salient points already arrived at, and the reasons that led to the next steps in our investigation. At a portion of the outer trench, it may be remembered, there was found, about a foot under the surface, a rude wooden platform resting on a complete solid basis, which then, naturally enough, was supposed to be the surface of the artificial island ; and towards the centre a series of at least four hearths, one above the other. Now the level of the lowest hearth was about 3 feet below that of the wooden platform. What then was the cause of this difference in their level ? Did the central portion sink from the weight of the superincumbent mass, or was it originally constructed so ? Again, although the fireplaces were nearly equidistant from the trench, measuring east and west (about 39 feet), they were eccentric in the diameter at right angles to this line, being, according to the measurements already given, about 14 feet north of the centre of the space enclosed by the trench. It was therefore evident that nothing short of the removal of a large portion of the central débris would be sufficient to give a correct idea of the log pavement and its surrounding structures, and disclose the treasures supposed to be hidden in it. Having adopted this resolution, the men were instructed accordingly, and at once commenced excavating directly south of the fireplaces. Part of the soil was thrown back into the empty space where the fireplaces stood, and the rest wheeled into the field beyond. The space thus inspected was about 25 feet broad, and extended southwards 31 feet from the fireplace. At its southern end we came upon a curved row of upright piles, most of which had the appearance of being dressed like square-cut beams, which penetrated deeply below the log pavement, and appeared to bound it in this direction. Amongst the relics found here were a pair of querns, portions of a wooden plate (Fig. 103), curious wooden implements (Figs. 118 and 119), a wooden hoe lying immediately above the log pavement, and close beside it some black vegetable substance like hair, and a few bone and horn implements. At its south-east corner we just touched the edge of a thick bed of ashes and bones, which will be described fully by and by.

We next removed a broad slice from the portion left standing to the west of the fireplace, and in consequence of certain peculiarities in the arrangement of numerous piles and horizontal beams observed at the north-west corner (see Fig. 40), we determined to remove altogether the broad ring now left between the outer trench and the space cleared in the interior.

It would be rather tedious to describe the various details of this work minutely ; besides, it is not necessary in order to convey a general idea of the results obtained. It was a work of many weeks, of great toil and labour, and of much and varied comment by outsiders. One or two visits to the crannog seemed to satisfy the curiosity of most people. There were, however, a few gentlemen whose enthusiasm

never fagged, amongst whom I have specially to mention Mr.
James Blackwood, F.S.A. Scot., who by constant attendance
and counsel rendered valuable aid in the successful accom-
plishment of these excavations. It will therefore be more
convenient to arrange the further observations I have to make

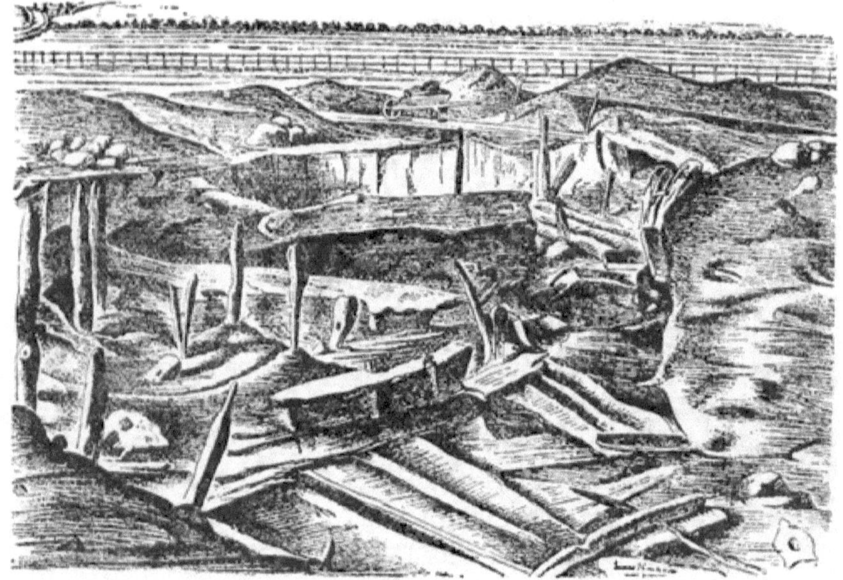

Fig. 40.—View of Wood-work at north-west corner.

in detailing the progress of the excavations under the
following heads :—

    1. Log pavement and its surrounding wooden structures.
    2. Ash and bone refuse-bed.

    1. *Log Pavement and its surrounding Wooden Structures.*—
After clearing the whole space enclosed by the original circular trench down to the level of the log pavement, it was
still difficult to make out the general plan of its structure,
and that of the superstructure erected upon it. In the
centre there was a rectangular space about 39 feet square,

having its sides nearly facing the four cardinal points, and a flooring of thick oak beams somewhat like railway sleepers (see Figs. 40, 41, and 42). The fireplaces were nearly in the centre, but a little nearer its northern side. The wooden pavement was more carefully constructed at the south side

Fig. 41.—Showing Horizontal Beam in its original position.

than under the fireplaces; although, quite close to the latter, on its eastern side, were found two beautiful slabs of oak, which were removed, and measured 12 feet by 1 foot 6 inches. These beams had a series of round holes extending along the whole length of one edge, and about $5\frac{1}{2}$ inches apart. They appeared quite symmetrical, as if formed by an auger, and had a diameter of about 1 inch, and a depth of 2 or 3 inches. Close to the southern side of this rectangular space, there were exposed two very curious beams 7 feet 9 inches apart, and lying over a thin layer of clay which intervened between them and the general log pavement.

One was slightly curved, and both had a raised rim running along their whole length, and each had a horizontal hole through which the ends of a beam passed (see Fig. 42). Moreover, they had square-cut holes at right angles to the former, as if intended for uprights. The finding of a double-bladed paddle (Fig. 126), close to one of these beams, suggested to the men the idea that they were the remains of a large boat, which, I must say, they very much resembled. Below this clay, and lying immediately over the log pave-

Fig. 42.—Curious Beams lying over Log Pavement.

ment, a long piece of a charred beam and the blade-half of an oar were found.

At the south-east and south-west corners of the wooden pavement the remains of what appeared to be partitions or walls, running northwards, were noticed (see Figs. 40 and 42). These were constructed of short uprights and long slender beams laid along the line of partition, and interspersed with a matty substance like bast, together with clay and earthy matter. At the south end, the logs forming the pavement were laid parallel to each other and in groups, some running north and south, and others at right angles to these. There were two and sometimes three layers of logs, each lying transversely over the other. At the ends of the upper layers there were here and there deeply penetrating piles slightly projecting above the flooring, with a horizontal beam stretched between and tightly jammed, apparently for the purpose of keeping the logs in position. About 12 or 13 feet from the south side, a straight row of these piles and stretchers ran across the log pavement, which, at first sight, I took to be the remains of a partition (see Plan of Crannog, Plate II.).

Surrounding the rectangular log pavement, and just touching its four corners, we could trace a complete circle of firmly-fixed upright piles, arranged in two rows from 2 to 3 feet apart. They were all made of oak. apparently young trees, and projected several feet above the surface of the pavement, some of which were observed on the grassy sur-face of the mound before excavations were commenced. The most important thing, however, about them was the mode in which they were connected together by transverse beams, similar to, but ruder than, those already described as found at the north-east corner of the outer trench. Some of these beams were bevelled at the ends on their upper sur-faces, especially the outer ends, and had two holes, one at each end, through which the pointed ends of the uprights projected. Fig. 41 shows one in its original position. At its inner end there were two strong wooden pins in a slanting direction, which entered the mortised hole through lateral grooves on its under surface and jammed the upright. The ends of these pins diverged and rested on clay, stones, and pieces of wood, and were evidently inserted for the purpose of supporting it. One transverse beam, observed on the west side not far from the former, and forming part of the same

elevated platform, had horizontal holes, and lay on a solid
mass of wood, stones, and vegetable matter, which was inter-
posed between it and the rude log pavement (the rectangular
oak pavement did not extend so far). Fig. 40 is a view
taken from about the middle of the bank, close to the south
side of the log pavement, and looking north-west. In front
are seen the remains of a partition, a little farther back the
beam just described, and turning round, at the far-off corner,
the beam represented in Fig. 41. Fig. 42 is also taken from
the same point, but with the view looking north-east. In
both these sketches portions of the oak pavement are seen
before any of the logs were disturbed. All the raised beams
found in position were from 2½ to 3 feet above the log pavement, and were directed towards the centre of the crannog,
so that they presented an appearance which reminded one of
the spokes of a large wheel. On the north side this arrangement was very well marked, many of the beams being still
in situ, and in one place long beams were found lying over
them, and running along the circumference of the crannog,
above which were distinctly seen remains of a wooden platform precisely similar to that already described, at the north-
east corner, with which, indeed, it was continuous.

It is thus more than probable that a circular platform of
wood, presenting a breastwork some 3 feet high, surrounded
the central log pavement, except at its southern side, where
no traces of the raised horizontal beams were found, and
where also the uprights were mostly formed of thick boards,
suggesting rather the idea of a division between the wooden
pavement and the refuse-bed. On the west side the segment left between the side of the rectangular oak pavement
was also covered with logs of wood, but much rougher, and
made of a softer wood than oak. This ruder pavement
extended below the transverse beams, and merged into a
conglomerated mass of stones, brushwood, and beams.

External to this circle of piles and platform, at the sides, but more especially on the south, there were other piles which appeared to form circles. On the south side indications of two or three such circles were noticed, but on the north side we could not ascertain their extent, as the trench was not far enough out to expose them if they did exist. But this point, together with several others, we hope to determine by further excavations as soon as the weather permits.

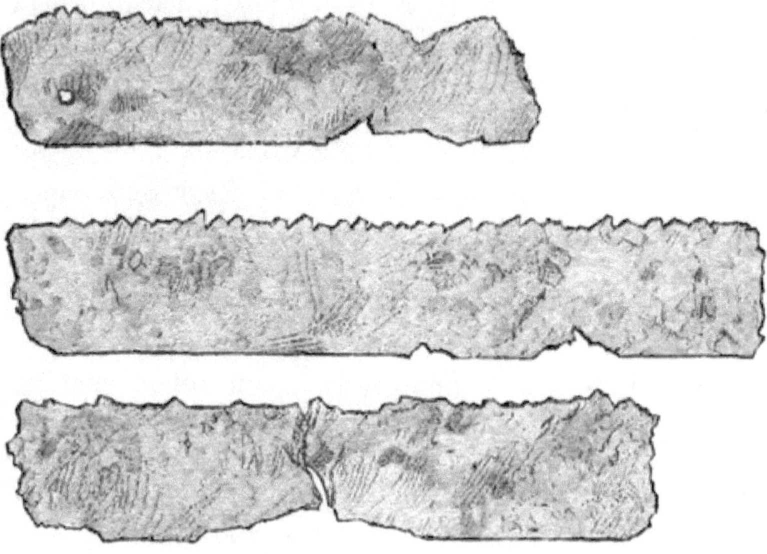

Fig. 43.—Portions of Iron Saw (¼).

About 25 yards south of the crannog I observed a row of stakes in an open drain running towards the nearest land, and the tops of others in the grass, which from their arrangement suggested the idea that they were part of a gangway which formerly extended between it and the shore. This is one of those points not examined when our operations were interrupted by the severity of the weather.

The principal relics found beyond the inner circle row

of piles consist of portions of a metal saw (Fig. 43), three
flint implements (Figs. 63, 64, and 65), and two bundles of
the fringe-like apparatus made of moss, besides those found
in the refuse-bed.

2. *Refuse-Bed.*—The refuse-bed lay at the south-east side
of the crannog (see Plan), just at the corner of the central
log pavement, and consisted chiefly of gritty ash, decayed
bones, and vegetable matters. It extended from the inner
circle of stockades to within a few feet of the outer trench.
Its breadth would be about 10 or 12 feet, and its length
from east to west nearly double that. Its surface was from
3 to 4 feet below that of the field, so that its average depth
would not be much short of 3 feet. Some important relics
were found here, such as metal instruments and daggers,
two fibulæ, several wooden vessels, and a few bone implements. It is noteworthy that the metal objects were all
comparatively near the surface of the midden, and also that
no boars' tusks or teeth were found in it except at its very
lowest stratum.

It was ascertained, through the careful inspection of the
Rev. Mr. Landsborough, that some of the large bones, especially leg-bones, contained in their cavities and interstices
beautiful green crystals, of which I have collected some fine
specimens. According to the analysis of Mr. John Borland,
F.C.S., F.R.M.S., they are Vivianite, regarding which he
writes as follows :—

" *Vivianite.*—A phosphate of iron, of somewhat definite composition, arising from the varying degree of oxidation of its base
and state of hydration.

"It is found in two conditions—Amorphous and Crystalline—
the former not uncommon, the latter rare. The amorphous has
been frequently described under the name of blue iron earth ; the
crystalline was first named, and its relationship to the amorphous
pointed out, by Weiner in Hoffmann's *Mineralogie*, about the year
1818 or 1820 ; the name being given in compliment to a Mr

Vivian of Cornwall, whose attention was first directed to the mineral.

"It has also been found at Bodenmais in Berne, and in several localities in America.

"Bischoff, in his *Elements of Chemical and Physical Geology*, as translated for the Cavendish Society, vol. ii. page 35, refers to a paper communicated by Von Carnall to a meeting of the Niederrheinischen Gesellschaft at Bonn, on the 3d December 1846, wherein mention is made of a remarkable instance of the occurrence of this mineral in the Scharley calamine mine, Silesia, which it was presumed was originally worked for lead.

"At a depth of 8 or 9 fathoms the skeleton of a man was found, and on breaking one of the bones crystals of vivianite became visible in the interior. A thigh-bone, when sawn through, showed crystals projecting from the inner surface, and others which were loose. The length of time the bones had lain there was unknown. The working of the Scharley mine began in the thirteenth century, and at the date of the communication had been discontinued for nearly three hundred years.

"Bischoff, however, advances the suggestion that, as the shaft may have been sunk in search of calamine, and not for the working of the lead, the age of the bones would not be so great as might at first be assumed. An analysis of the few crystals placed at my disposal leads to the conclusion that their constitution may be represented by the formula

$$3FeO.P^2O^5 + Fe^2O^3.P^2O^5 + 15 \text{ aq.}$$

They belong to the monoclinic system of crystallography, and are of greenish-blue colour, becoming darker gradually on exposure to air."

In several places, when digging below the level of the log pavement and thrusting a staff a few feet downwards, gas bubbled up through the water, which, on applying a lighted match, ignited with considerable explosion. This, on analysis, was found to be carburetted hydrogen or marsh gas, with a small quantity of carbonic acid gas.

Before the stuff inside the circular trench was completely cleared away down to the level of the log pavement, our operations had to be abandoned on account of the severity of the weather. Meantime I drew up the above report from a careful journal kept of each day's proceedings and finds, and at the March meeting communicated it to the Society of Antiquaries of Scotland. But, notwithstanding the great variety of relics discovered, and the important information regarding the general structure of the crannog which had been ascertained, there were still several points requiring further elucidation. Of these the following four were the chief, which may be thus succinctly stated:—

Firstly.—From a perusal of the Plan (Plate II.) it will be observed that at the south side there is at least one well-marked circular group of upright piles external to the one surrounding the log pavement; hence the question which pressed for solution was—Whether these groups merged into the one on the north side, or whether there was another corresponding to the former still further out?

Secondly.—It was obvious that the island extended considerably beyond our original circular trench, so that a correct estimate of it could not be formed from our present data.

Thirdly.—We had no reliable information regarding the composition of the island below the log pavement, as deeper digging could not be carried on to any extent without a pump, owing to the accumulation of water—the main drain being nearly on a level with it.

Fourthly.—The supposed gangway had to be examined.

As none of the above problems could be solved without additional excavations, it was clear that, in the interests of science, the work should be resumed. But here occurred a difficulty. As the drainage operations conducted on the farm of Lochlee had now come to a close, and the workmen were removed elsewhere, Mr. Turner gave instructions that

no further outlay should be incurred in the investigation of
the crannog; and as, moreover, his Grace the late Duke of
Portland, in answer to petitions from the Town Council and
Philosophical Society of Kilmarnock, had given all the relics
to the Corporation of this town, we felt it incumbent on us
to restrict applications for more funds to carry on the explorations to the local authorities who had thus, without any
expenditure whatever, become the owners of a rare and
valuable collection of archæological relics.   But the only
result of our representation was a grant of £10 from the
Philosophical Society; which, however, under the judicious
management of Mr. Blackwood, together with a few private
contributions kindly given by Messrs. James Blackwood,
James Craig, Charles Reid, and Thomas Kennedy, enabled
us to bring the work to a tolerably satisfactory conclusion.

Upon resuming operations in the month of April we
directed the workmen to clear away the soil at the north-
west corner, where, it will be remembered, two mortised
beams were exposed in the original circular trench.   These
were then supposed to be part of the well-defined circle
running along the north side, but now, however, they were
found to be from 8 to 10 feet external to this circle.   Upon
careful inspection of the wooden structures at the north-east
corner, we found that the inner termination of the platform,
conterminous with the elaborate mortised beams at the outer
trench, was supported by transverse mortised beams similar
to those in the general circle—one of which is figured in Fig.
41.   There could, indeed, be hardly any doubt that at this
corner two circular rows of uprights with their transverses
gradually merged into one on the north.   Hence it became
a very feasible supposition that those mortised beams at the
north-west corresponded with the outer ones at the north-
east side, and formed part of an outer circle which also

merged into the one on the north. But upon extending excavations so as to expose them completely, this supposition was not borne out. They were in a slanting position, about 15 feet apart, and their outer ends on a level with the log pavement. Half-way between them there was another beam lying in a similar position, but it contained no mortised holes. Their lower or outer extremities were jammed against a sort of network of logs, some running along the circumference and others slanting rapidly downwards, while their inner ends were raised about 2 feet and rested on a mass of stones and logs of wood. The outer hole of the beam, marked H on the Plan, contained a portion of an upright, which had, however, more the appearance of being used as a peg to keep it down. The other mortised holes appeared to be of no use whatever, so that these beams were intended for, and probably served, a different purpose before being placed in their present position.

It was now evident that the margin of the crannog was near, as at the upper or surface portions of the trenches we encountered a layer of fatty clay, which had undoubtedly been deposited by the surrounding lake. This layer gradually got thicker as we advanced outwards, and the dark vegetable débris and wood-work, forming the substance of the island, shelved downwards underneath it. A foot or two beyond the outer end of the beam G, this clay was 3 feet 6 inches thick. Pursuing our investigations northwards towards the point A (Plan Plate II.), we came upon a dense wooden structure formed of stakes, logs, planks, and brush-wood, woven together in the most fantastic fashion, which also shelved downwards below the clay. At the point A this clay was no less than six feet deep. Here the water oozed up, but there was no doubt, from the above appearances and the rapidly slanting wood-work,—some stakes now running downwards and outwards at an angle of about

45°,—that we had reached the sloping margin of the island. Imbedded in the clay near the point A were found two pieces of charred stakes, one 3½ feet and the other nearly 6 feet deep. About half-way between the margin of the crannog and the circle of stakes surrounding the log pavement, and 5 feet deep, the workmen discovered, amongst decayed brushwood and chips of wood, a beautiful trough cut out of a single block of wood. It was quite whole when found, and showed very distinctly the markings of the gouge-like instrument by which it was fashioned. It was made of soft wood, which, upon drying, quickly crumbled into dust, but Fig. 44, engraved from a photograph taken by Mr. Blackwood soon after its discovery, gives a very good idea of it.

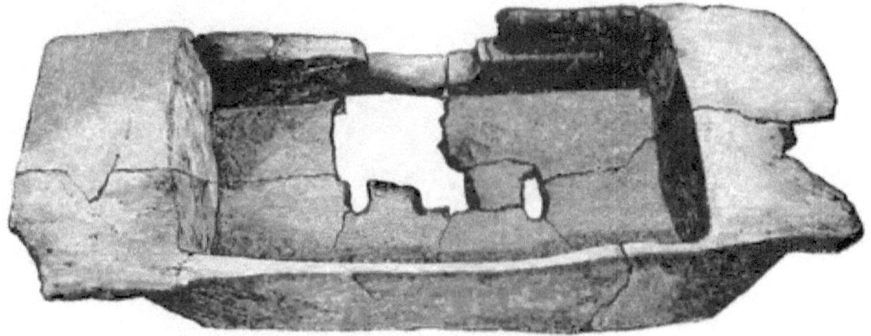

Fig. 44.—Wooden Vessel (¼).

Instead of pursuing the excavations further in this direction, our means being quite inadequate to clear away the soil at a uniform breadth of about 20 feet all round, we resolved to form a number of cuttings projecting outwards, at suitable intervals, from the circumference of the space already cleared. These cuttings (see Plan, A, B, C, D and E) varied from 10 to 20 feet in breadth, and extended outwards in each case till we were satisfied, from the encroachment of the surrounding clay, that the margin of the crannog had

been reached. On the north and north-east trenches the
wood-work assumed a most extraordinarily intricate arrangement. It consisted mostly of young trees and branches of
birch, the bark of which was quite fresh-like, and distinctly
recognisable, mixed with stakes and logs, some of oak, running in all conceivable directions, and constituting a protective barrier,—proof, I should say, against the most violent
action of both wind and water. At its inner side, close to
the original circular trench, this peculiar structure, which
we called trestle-work, was only about 18 inches below the
surface, but sloped downwards, at first gradually, and then
rapidly, till it disappeared under the clay. At the north-
east corner it extended about 20 feet beyond the group of

Fig. 45.—Wooden Board (¼).

mortised beams, so that the latter could not have been a
landing-stage, a theory which was long current amongst the
quidnuncs. Near the outer edge of the cutting at this
corner (c), there was observed, mixed up with the trestle-
work, an oak beam, having two square mortised holes, which
must have been originally adapted for a higher purpose than
the humble function of packing, which it here served.
Lying over the wood-work, and less than two feet below the
surface, I picked up portions of a leather boot or shoe, with
fragments of a leather lace, crossed diagonally, which had
tied it in front; also a small wooden stave like that of a
milk-cog. Deeper, and near the outer edge, the workmen
found a much corroded dagger or spear head. At the south-

east corner (D), a series of upright piles with the remains of a transverse was exposed, but the trestling work had dwindled down to mere brushwood, with an occasional beam mixed up with it. Here the workmen found a thin board made of hard wood, resembling a portion of the end of a small barrel, with diagonal and other markings lightly cut upon it (Fig. 45).

On the south side, external to the refuse-bed, quite a forest of piles was encountered, together with the charred remains of a few mortised transverses and some long beams. From a glance at the Plan it will be observed that, at the cutting E, the outer circle of these uprights curves outwards as if to meet the line of the supposed gangway. It would have been more satisfactory if a larger portion had been here cleared away, and the junction of the gangway with the crannog more accurately determined; but at this particular spot there was such an immense accumulation of rubbish, formerly wheeled from the interior of the mound, that the labour of removing it was too great. The superficial layer of fatty clay appeared here also, and at the point E measured 2 feet 3 inches in thickness. The horizontal beams found at this side, some of which were indicated on the Plan, were from 4 to 5 feet deep, and about the same level some important relics were dug up. Near the point M were found a bridle bit (Fig. 148), a bronze dagger-like instrument (Fig. 145), and a four-plied plaited object made of the long stems of a moss similar to those of which the fringe-like article was manufactured, and referred to on a former occasion. It had the tapering appearance of a cue or pigtail; and measured 17 inches long and about 2 broad in the middle. Near it, and about 5 feet deep, an iron hatchet (Fig. 46), much corroded, but still retaining a small bit of the wooden handle, was discovered by one of the workmen. A few feet to the east of this, and lying across the line of the gangway, a large oar was exposed to view.   It was quite whole when found, but being made of soft wood, was so fragile that it broke into pieces in the act of removal.   Its extreme length was 9½ feet,

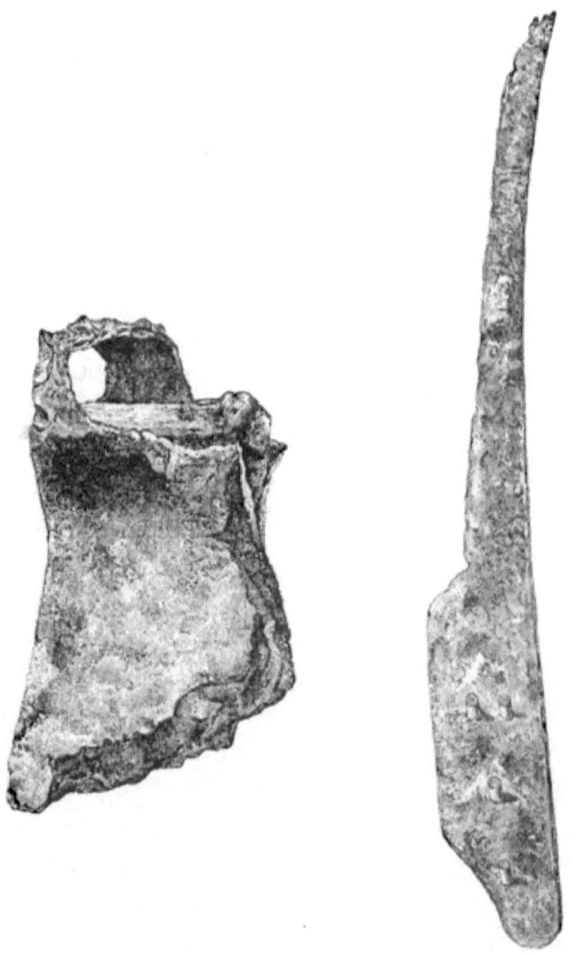

Fig. 46.—Iron Hatchet (⅓).   Fig. 47.—Iron Knife (⅓).

and the blade measured 3 feet by 14 inches.   The round handle was perforated about its middle by two small holes a couple of inches apart.

We made no projecting trench on the south-west side

owing to the proximity of a network of recent drains, which, if disturbed, might injuriously interfere with the drainage of the field, but from the general appearance of the wood-work we were satisfied that this portion was symmetrical with the rest of the crannog. The ends of flat beams jutted out at the bottom of the cutting immediately on the west side, which clearly indicated a parallelism with the three exposed a little farther north; and towards the south one or two uprights belonging to the outer series were visible.

Having now collected the chief facts regarding the log pavement, its surrounding and superincumbent structures, and the extent of the crannog, we determined to sink a shaft at the lower end of the log pavement—*i.e.* about the centre of the crannog—for the purpose of ascertaining, if possible, the thickness, composition, and mode of structure of the island itself. This shaft was rectangular in form, and large enough to allow three men to work in it together. After removing the three or four layers of oak planks which constituted the log pavement, we came upon a thin layer of brushwood and then large trunks of trees laid in regular beds or layers, each layer having its logs lying parallel to each other, but transversely and sometimes obliquely to those of the layer immediately above or below it. At the west end of the trench, after removing the first and second layers of the log pavement, we found part of a small canoe hollowed out of an oak trunk. This portion was 5 feet long, 12 inches deep, and 14 inches broad at the stern, but widened towards the broken end, where its breadth was 19 inches. This was evidently part of an old worn-out canoe, thus economised, and used instead of a prepared log. Much progress in this kind of excavation was by no means an easy task, as it was necessary to keep two men constantly pumping the water which copiously flowed from all directions into the trench, and even then there always remained some at the bottom.

As we advanced downwards we encountered layer upon layer of the trunks of trees with the branches closely chopped off, and so soft that the spade easily cut through them. Birch was the prevailing kind of wood, but occasionally beams of oak were found, with holes at their extremities, through which pins of oak penetrated into other holes in the logs beneath. One such pin, some 3 or 4 inches in diameter, was found to pass through no less than four beams in successive layers, and to terminate ultimately in a round trunk over 13 inches in diameter. One of the oak beams was extracted entire, and measured 8 feet 3 inches in length and 10 inches in breadth, and the holes in it were 5 feet apart. Others were found to have small round projections, which evidently fitted into mortised holes in adjacent beams.

Down to a depth of about 4 feet the logs were rudely split, but below this they appeared to be round rough trunks, with the bark still adhering to them. Their average diameter would be from 6 inches to 1 foot, and amongst them were some curiously gnarled stems occasionally displaying large knotty protuberances. Of course the wood in the act of digging the trench was cut up into fragments, and, on being uncovered, its tissues had a natural and even fresh-like appearance, but in a few minutes after exposure to the air they became as black as ink. Amongst the débris thrown up from a depth of 6 feet below the log pavement I picked up the larger portion of a broken hammer-stone or polisher, which, from the worn appearance presented by its fractured edges, must have been used subsequently to its breakage. After a long and hard day's work we reached a depth of 7 feet 4 inches, but yet there were no indications of approaching the bottom of this subaqueous fabric. However, towards the close of the second day's labour, when the probability of total discomfiture in reaching the bottom was freely talked of, our most energetic foreman announced, after cutting through a large flat trunk 14 inches thick, that underneath this he could find no trace of further wood-work. The substance removed from below the lowest logs consisted of a few twigs of hazel brushwood, imbedded in a dark, firm, but friable and somewhat peaty soil, which we concluded to be the silt of the lake deposited before the foundations of the crannog were laid. The depth of this solid mass of wood-work, measuring from the surface of the log pavement, was 9 feet 10 inches, or about 16 feet from the surface of the field.

Amongst the very last spadefuls pitched from this depth was found nearly one-half of a well-formed and polished ring made out of shale, the external and internal diameters of which were 3½ and 2 inches respectively.

Gangway.—The probable existence of some kind of communication between the crannog and the shore of the lake was suggested at a very early stage of these investigations by the discovery of a few oak piles in a drain outside the mound, and to clear up this mystery was now the only problem of importance that remained to be solved. We commenced this inquiry by excavating a rectangular space, 30 feet long, 16 feet broad, and 3 to 4 feet deep, in the line of direction indicated by the piles (marked o on the Plan), and exposed quite a forest of oak stakes. Other trenches, marked P and Q respectively, were then made with exactly similar results. The stakes thus revealed did not at first appear to conform to any systematic arrangement, but by and by we detected, in addition to single piles, small groups of three, four, and five, here and there at short intervals. This observation, however, conveyed little or no meaning, so that we could form no opinion as to the manner in which they were used. No trace of mortised beams was anywhere to be seen. In all the trenches the stuff dug up was of the same character. First or uppermost there was a bed of fine clay rather more than 2 feet thick, and then a soft dark substance formed of decomposed vegetable matters. The source of the latter was evident from the occurrence in its upper stratum of large quantities of leaves, some stems, branches, and the roots of stunted trees, apparently *in situ*. The tops of the piles in the trench Q were from 2 to 3 feet below the surface of the field, but they appeared to rise gradually as we receded from the crannog, and in the trench next the shore one or two were found on a level with the grass. About 4 feet deep the stuff at the bottom of the trench was so soft that a man could scarcely stand on it without sinking ankle-deep. It was not nearly so heavy as ordinary soil, but more adhesive, and of a nutty brown colour, which, on exposure, quickly turned dark. Notwithstanding the flabbiness of this material, the piles felt quite firm, and this fact, together with the experience derived from our examination of the deeper structures of the island, led to the supposition that the piles would be found to terminate in some more solid basis than had yet been made apparent. To remove all doubts on this point, though a long iron rod could be easily pushed downwards without meeting any resistance, we ordered a large deep shaft to be dug in the line of the piles, and the cutting Q, being nearest the crannog, was selected for this purpose. This was accomplished with much difficulty, but we were amply rewarded by coming upon an elaborate system of wood-work, which I found no less difficult to comprehend than it now is to describe. The first horizontal beam was reached about 7 feet deep, and for other 3 feet we passed through a complete network of similar beams, lying in various directions. Below this, *i.e.* 10 feet from the surface, the workmen could find no more beams, and the lake silt became harder and more friable. We then cleared a larger area so as to exhibit the structural arrangement of the wood-work.   The reason of grouping the piles now became apparent.   The groups were placed in a somewhat zigzag fashion near the sides of the gangway, and from each there radiated a series of horizontal beams, the ends of which crossed each other and were kept in position by the uprights.   One group was carefully inspected.   The first or lowest beam observed was right across, the next lay lengthways and of course at right angles to the former, then three or four spread out diagonally, like a fan, and terminated in other groups at the opposite side of the gangway, and lastly, one again lay lengthways.   (See Plan and Sections.)   Thus each beam raised the level of the general structure the exact height of its thickness, though large lozenge-shaped spaces remained in the middle quite clear of any beams.   The general breadth of the portion of this unique structure examined was about 10 feet (but an isolated pile was noticed farther out), and its thickness varied from 3 to 4 feet.   A large oak plank, some 10 feet long, showing the marks of a sharp cutting instrument by which it was formed, was found lying on edge at its west side, and beyond the line of piles, but otherwise no remains of a platform were seen.   All the beams and stakes were made of oak, and so thoroughly bound together that, though not a single joint, mortise, or pin was discovered, the whole fabric was as firm as a rock.   No relics were found in any of the excavations along the line of this gangway.

RELICS.

The remains of human industry found during the excavations of the Lochlee Crannog, calculated to throw light on the civilisation and social economy of its occupiers, are very abundant.   They comprise a large variety of objects, such as warlike weapons, industrial implements, and personal ornaments, made of stone, bone, horn, wood, metal,

etc. In the following description of them I have adopted,
as perhaps the most convenient, the principle of classification suggested by the materials of which they are composed.

I. OBJECTS MADE OF STONE.

Hammer-Stones.—A great many water-worn pebbles, of
a similar character to those found in the surrounding glacial
drift and river-courses, which were used as hammers, or
pounders, or rubbers, were discovered in the débris all over

Fig. 48.—Hammer-Stone (½).   Fig 49.—Hammer-Stone (½).

the crannog, but more abundantly in the deeper layers of a
small circular area round the hearths, corresponding to what
I have on a former occasion designated the relic-bed. As
typical specimens of such implements I have collected no
less than nineteen. Of these, fourteen are of a somewhat
elongated oval shape, and were used at one or both ends.
They vary considerably in size, the major diameter of the
largest measuring 6 inches, and the rest graduating downwards to about the half of this. Two are flat and circular,
and show friction-markings all round; while other three
show signs of having been used on their flat surfaces only.
The one represented in Fig. 48, with markings on its flat

sides, is divided into two portions, each of which was picked up separately, about a yard asunder, and found to fit exactly. It would thus appear that it was broken while being used on the crannog, and then pitched aside as useless. Some are slightly chipped at one end, others have small finger-like depressions, as if intended to give the user a better grip (Figs. 49, 50, and 51).

Fig. 50.—Hammer-Stone (½).

Fig. 51.—Hammer-Stone. Edge view of the previous implement (½).

Heating-Stones and Sling-Stones.—A large number of round stones, varying in size from half an inch to three inches in diameter, some having their surfaces roughened and cracked as if by fire, but others presenting no marks whatever, were met with. The former might have been used as heating-stones for boiling water in wooden vessels,—the only ones found on the crannog,—the latter as sling-stones or missiles.

Anvil.—About a foot below the surface, and a few feet to the north of the upper fireplace, a beautiful quartz pebble was found by Mr. Cochran-Patrick, which has the appearance of being used as an anvil. It is discoidal in shape, but

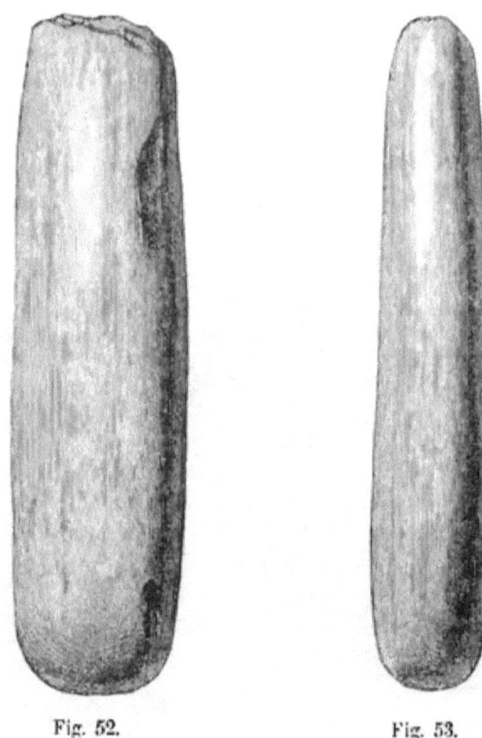

Fig. 52.   Fig. 53.

Sharpening-Stones ($\frac{1}{4}$).

a little more rounded on its upper surface, and measures 27 inches in circumference. It is just such an instrument as a shoemaker of the present day would gladly pick up for hammering leather (see Fig. 37).

Sharpening-Stones or Whetstones.—Four or five whetstones were collected from various parts of the island, three of which are here engraved (Figs. 52, 53, and 54).   They are made of a hard smooth claystone, one only being made of a fine-grained sandstone, and vary in length from 5 to 7 inches. Fig. 54 represents what is supposed to be a hone 6¼ inches long, and containing a smooth groove.   It was found on the site of the crannog by Captain Gillon, long after the explorations had been brought to a close, and is now deposited in the National Museum (see page 126).

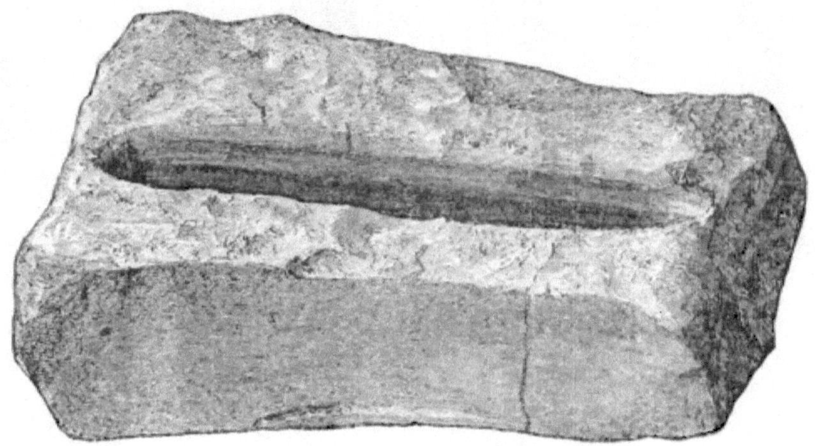

Fig. 54.—Hone (6¼ inches in length).

Besides these *hones* we noticed a large block of a coarse sandstone, having one side covered with deep ruts, supposed to be caused by the sharpening of pointed instruments.

Polished Celt.—Only one polished stone celt was found. It is a wedge-shaped instrument, 5½ inches long, and 2 broad along its cutting edge, which bears the evidence of having been well used, and tapers gently towards the other end, which is round and blunt.   It is made of a hard mottled greenstone (Fig. 55).

Circular Stone.—Fig. 56 represents a peculiar circular implement manufactured out of a bit of hard trap rock.

It presents two flat surfaces, 3 inches in diameter, with a round periphery, and is $1\frac{3}{8}$ inch thick.

Fig. 55.—Stone Celt ($\frac{1}{2}$).

Querns.—Five upper, and portions of several lower, quern-stones were disinterred at different periods during

Fig. 56.—Circular Stone ($\frac{1}{2}$).

these excavations, all of which, however—with the exception of the pair found over the log pavement, and an upper stone observed towards the west margin of the crannog, but of which I could find no definite information, as it was stolen soon afterwards,—were imbedded in the débris not far from the site of the fireplaces, and superficial to the level of the middle or first-discovered pavement.   Some are made of

Fig. 57.—Upper Quern Stone (⅛).

Fig. 58.—Upper Quern Stone (⅛).

granite, while others appear to be made of schist or hard whinstone.   Besides the central cup-shaped hole, which, of course, all the upper ones possess, one has a second hole slanting slightly inwards, another has a similar hole, but only half-way through, while a third has no second hole at all, and a fourth shows a horizontal depression at its side. The one without a second hole on its surface is nearly circular, but the others are all more or less elongated.   Their

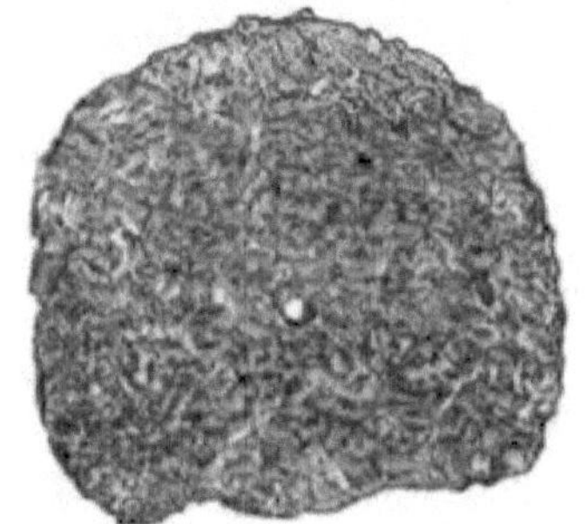

Fig. 59.—Lower Quern-Stone (⅛).

largest diameters vary from 13 to 14 inches. One is broken into three portions, which, though dug up separately, fit exactly. It measures 14 inches by 11, and the central hole is wide, being no less than 5 inches across. From the upper edge of this hopper-like cavity the stone slopes gently all round to the circumference of its under surface, and the second hole completely perforates it.

Cup-marked Stones.—Two portions of red sandstone, having cup-shaped cavities about 1 inch deep and 3 inches diameter, were found amongst the débris. One of them was lying underneath a horizontal raised beam at the north side of the crannog. The position of the other was not determined. The latter has two circular depressions or grooves round the cup, the outer of which is about 9 inches in diameter (Figs. 60 and 61).

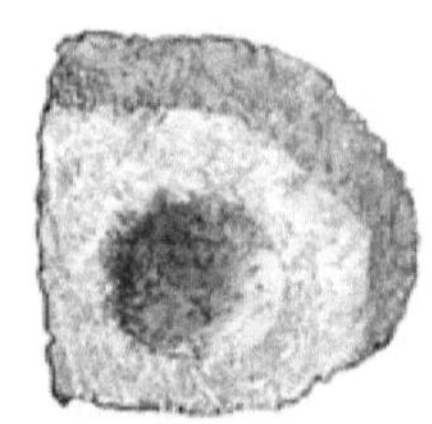
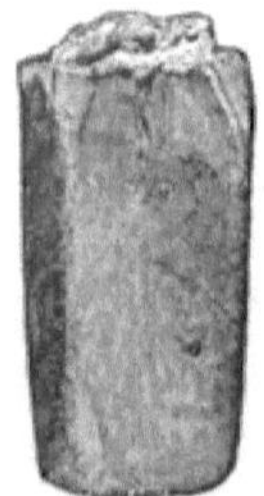

Fig. 60.—Cup Stone (¼).   Fig. 61.—Cup Stone (⅛).   Fig. 62.—Stone (½).

Other Stone Relics.—Besides the above there are a few other articles of stone bearing the evidence of design, which I must just allude to.

1. A large stone having a deep groove all round about it, as if intended for a rope. The larger portion of this groove was caused by atmospheric agencies, and only one side could be positively stated to have been artificially formed.

2. A thin oval-shaped disc of a light black substance like shale, measuring 3 inches by 2 inches.

3. Portion of a polished stone 2 inches long, having a narrow groove surrounding one end, and through which it appears to have been broken (Fig. 62).

Flint Implements.—Only three flint objects have been discovered on the crannog.

1. A beautifully chipped horseshoe-shaped scraper, found at north-east corner, on a level with the raised wooden platform. It is made of a whitish flint, and measures 1 inch in length by 1¼ in breadth (Fig. 63).

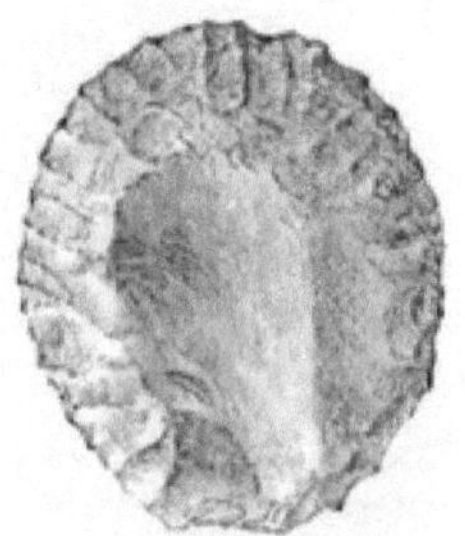

Fig. 63.—Flint Scraper (¼).

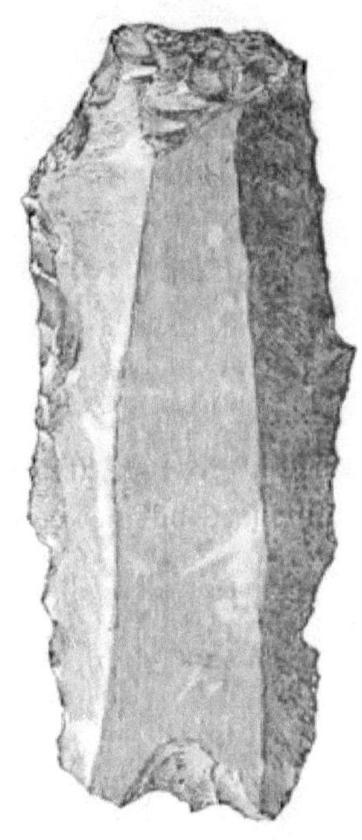

Fig. 65.—Posterior of Flint Flake (½).

Fig. 64.—Flint Flake (½).

2. A large knife-flake, 3 inches long and 1¼ broad, which appears to have been much used at the edges and point. It is also made of a whitish flint, and presents three smooth surfaces above and one below (Fig. 64).

3. The end portion of another flake, made of a dark flint (Fig. 65).

Spindle Whorls.—Three small circular objects, supposed to be spindle whorls, are here classed together. Two are made of clay, and were found in the relic-bed near the fireplaces. The smaller of the two is 1¼ inch in diameter, and has a small round hole in the centre; the other has a diameter of 1¾ inch, but is only partially perforated, just

Fig. 66.—Clay Spindle Whorl (½).

Fig. 67.—Clay Spindle Whorl (½).

sufficient to indicate that the act of perforation had been commenced but not completed (Figs. 66 and 67). The third object is a smooth, flat, circular bit of stone, 1½ inch in

Fig. 68.— Stone (½).

diameter and ½ an inch thick, and is perforated in the centre like a large bead (Fig. 68).

A fourth spindle whorl was found by a visitor from Dumfries-shire and carried away, but it has been returned lately, and is now among the collection. It is made of an oval sandstone pebble, and is slightly larger than the one represented by Fig. 68.

II. Objects of Bone.

Upwards of twenty implements made of bone have been added to the general collection, all of which were found either in the relic-bed or refuse-heap. The following are the most interesting :—

1. Two Chisels or Spatulæ. One (Fig. 69) is made of a split portion of a shank-bone, and measures 5¼ inches long and rather less than ½ an inch broad. It is very hard, flat, and smoothly ground at one end, and has a sharp rounded edge, which extends further on the left side, thus indicating that it was adapted for being used by the right hand. The other (Fig. 70) is a small leg-bone obliquely cut so as to present a smooth polished surface. Its length is 4 inches and diameter ½ inch.

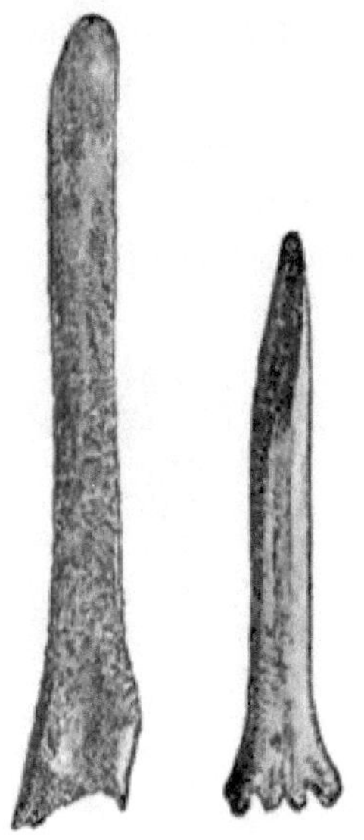

Fig. 69.   Fig. 70.
Bone Chisels (½).

2. Five small objects presenting cut and polished surfaces, three of which are sharp and pointed (Figs. 71, 74, 75); one (Fig. 72) appears to have been notched

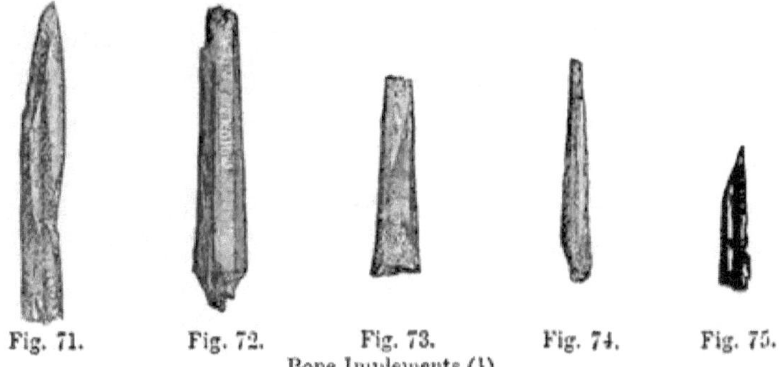

Fig. 71.   Fig. 72.   Fig. 73.   Fig. 74.   Fig. 75.
Bone Implements (½).

at the end and there broken off; and the last (Fig. 73) presenting well-cut facets, is fashioned into a neat little wedge.

3. Fig. 76 represents a tiny little spoon only ¾ of an inch in diameter, and worn into a hole in its centre. The handle portion is round and straight, and proportionally small, being only two inches long and about the diameter of a crow-quill. Fig. 77 shows another portion of bone somewhat spoon-shaped.

4. Fig. 78 is a drawing of a neatly formed needle-like instrument. It is flat on both sides, finely polished, and gradually tapering into points at its extremities. The eye

Fig. 76.—Bone (½).   Fig. 77.— Bone (½).   Fig. 78.—Bone (½).

is near its middle, being two inches from one end and 1½ inch from the other, and large enough for strong twine to pass through it.

5. Fig. 79 is a drawing of a portion of bone artificially made into a sharp-pointed instrument. Several similar objects were met with, but as they showed no distinct workmanship I have not preserved them.

6. A great many small ribs, about 6 or 7 inches in length, and portions of others, were found to have the marks of a sharp cutting instrument by which they were pointed and smoothed along their edges, the use of which can only be conjectured. Figs. 80 to 82 are drawings of some of them. Fig. 83 shows a larger rib-bone, highly polished all over and notched round one end.

7. Lastly, there are several portions of round bones which appeared to have been used as handles for knives or suchlike instruments.

III. Objects of Deer's Horn.

Out of about forty portions of horn, chiefly of the red deer, bearing evidence of human workmanship, I have

Fig. 79.—Bone (⅓).   Fig. 80.   Fig. 81.   Fig. 82.   Fig. 83.
Bone Implements (½).

selected for illustration sixteen of the most characteristic specimens. Two hammers or clubs, formed from the lower portions of the beam antlers of stags by cutting or sawing off their branches. One (Fig. 84) is 11 inches long, and has about three inches of the brow branch of the horn projecting from it, round the root of which there is a groove as if intended for a string. The markings on the back portion

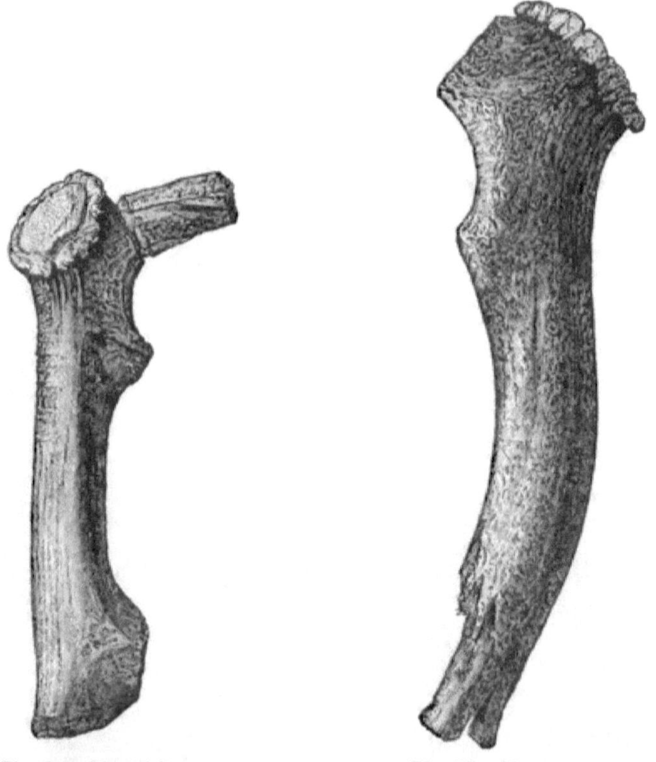

Fig. 84.—Horn (⅓).   Fig. 85.—Horn (⅓).

indicate very distinctly that it was used for hammering
some hard substance. Fig. 85 is a still more formidable

Fig. 86.—Horn (⅓).

Fig. 87.—Horn (⅓).

weapon, being 14 inches long and 9 inches in circumference near the burr. Portion of the latter is worn completely away by use. Fig. 86 is the root portion of a large antler,

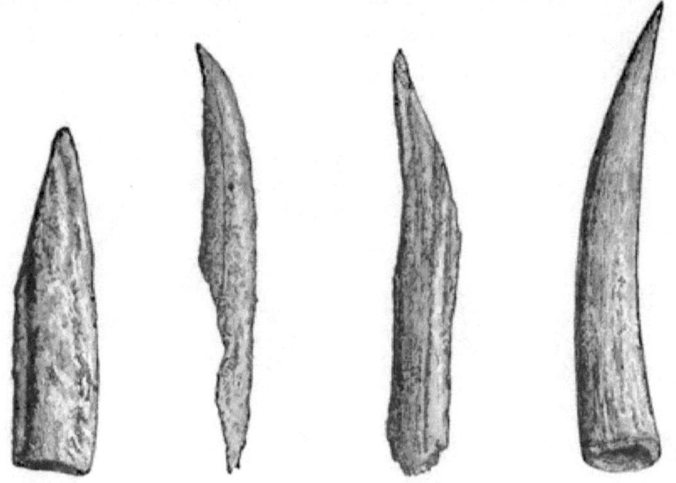

Fig. 88.—Horn (⅓).   Fig. 89.   Horn (⅓).   Fig. 90.—Horn (⅓).   Fig. 91.—Horn (⅓).

having one surface made smooth, and containing two circular depressions and a few deeply penetrating marks as if made by a sharp instrument.   Fig. 87 is a portion of a horn with

Fig. 92.—Horn (⅓).   Fig. 93.—Horn (⅓).   Fig. 94.—Horn (⅓).   Fig. 95.—Horn (⅓).

a groove round one end.   Figs. 88, 89, 90 represent split portions of horn sharpened at the point like daggers.   Figs.

91, 92, 93 are three pointed portions or tines, two of which were probably used as spear-heads, and contain small holes at the cut ends by which they were fastened on handles. Fig. 94 represents portion of horn (roe) cut at both ends with a hole near its centre, which, however, does not pass through ; while Fig. 95 shows another small pointed and

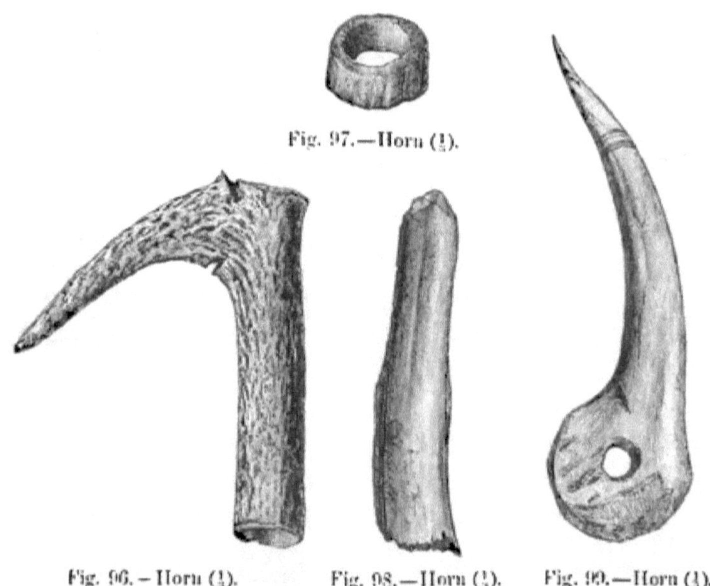

Fig. 97.—Horn (⅓).

Fig. 96.—Horn (⅓).   Fig. 98.—Horn (⅓).   Fig. 99.—Horn (⅓).

curved portion, with a hole, about 1 inch from the end, passing completely through it. Fig. 96 was evidently used as a hook, as the stem portion is smoothly bored and made

Fig. 100.—Portion of Horn Handle found along with Iron Knife, Fig. 129 (⅓).

suitable for a handle. Fig. 97 is a small portion made into a ring. The last object figured under this head is a *bodkin* 8 inches long, finely polished all over, and pointed at the tip as if with a sharp knife. The other end, which is large and circular, is pierced by a round hole, by means of which it might have been strung to one's person (Fig. 99). The portions of horns not figured consist of clubs, pointed tines, short thick pieces, etc., all of which show the marks of tools upon them.

Besides the above there are a great many fragments of horns, some of which, as mentioned by Professor Rolleston in his report on the fauna, might have been used as implements. One of the fragments labelled by this gentleman as being part of the horn of a reindeer, is a short flat tine, and bears the evidence of having been sawn off. It is 6 inches long and 2 broad at the base.

IV. OBJECTS OF WOOD.

A large assortment of wooden implements was found chiefly in the refuse-heap, and in the portion of débris corresponding to the area of the log pavement. Owing to the softness of the wood and the large amount of moisture contained in its fibres, most of these relics have already shrunk to less than half their original bulk, and become so changed, though they were kept in a solution of alum for several weeks, that I am doubtful of being able to preserve them at all. Seeing the rapid decay they were undergoing, I got full-sized pencil-drawings taken of them, from which the accompanying illustrations have been engraved. They consist of bowls, plates, ladles, a mallet, a hoe, clubs, pins, etc., together with many objects entirely new to me, but which apparently had been used for culinary or agricultural purposes.

1. *Vessels.*—Fig. 101. Portions of a circular bowl, diameter 7½ inches, depth (inside) 3 inches, thickness ¼ inch at edges and ½ inch at bottom; bottom flattened, 3 inches diameter (outside). Other fragments of vessels similar to the above were found.

Fig. 102. Flat dish, like scallop shell, with a ring handle, length 7 inches, breadth 6 inches, thickness varies from $\frac{3}{8}$ inch to a thin edge. Quite whole when disinterred from refuse-heap.

Fig. 103. Portions of a plate, diameter nearly 10 inches, thickness $\frac{3}{8}$ of an inch, depth barely 1 inch; a well-formed bead ran round the rim.

Fig. 104. Ladle. Bowl nearly complete, length 10 inches, breadth 8 inches, depth (inside) $3\frac{1}{2}$ inches, thickness 1 to $\frac{1}{2}$ inch; portion of handle still remaining.

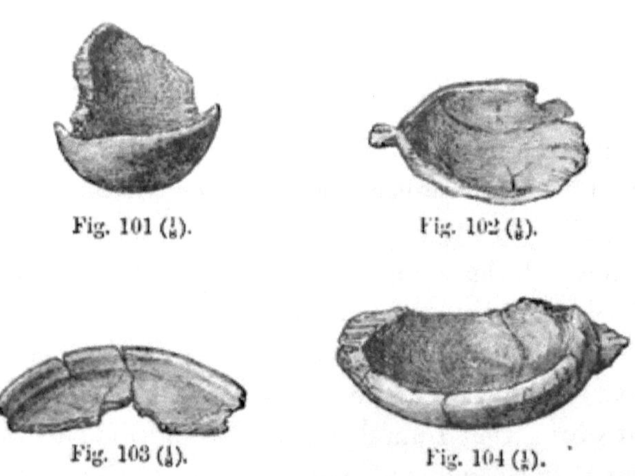

Fig. 101 ($\frac{1}{8}$).

Fig. 102 ($\frac{1}{8}$).

Fig. 103 ($\frac{1}{8}$).

Fig. 104 ($\frac{1}{8}$).

Fig. 44. Trough, $11\frac{1}{2}$ inches long, 6 broad, and $2\frac{1}{2}$ deep (inside). Projecting ears $3\frac{1}{2}$ inches long. Thickness of sides varied from $\frac{1}{4}$ to 1 inch. Had three rectangular holes in bottom, of which the centre one was larger, measuring 1 by $1\frac{1}{2}$ inch.

All the above vessels were made of soft wood, with the exception of the portions of bowls, which were of oak.

2. *Clubs, Pins, etc.; all of which were made of oak.*—Fig. 105. Club, 2 feet long, 3 inches broad, and $1\frac{1}{2}$ thick; circumference of handle $3\frac{1}{2}$ inches.

Fig. 106. Club, $14\frac{1}{2}$ inches long, and greatest breadth $2\frac{1}{2}$.

Fig. 107.   Sword-like implement, 20 inches long and 2½ broad; sharp at point and edges.

Fig. 108.   Implement with round handle and thin blade, containing teeth at one edge, length 15 inches, and breadth 1½.

Fig. 109.   Knife-shaped instrument, blade 10 inches long by 1 broad.

Fig. 110.   Round polished stick with charred end.

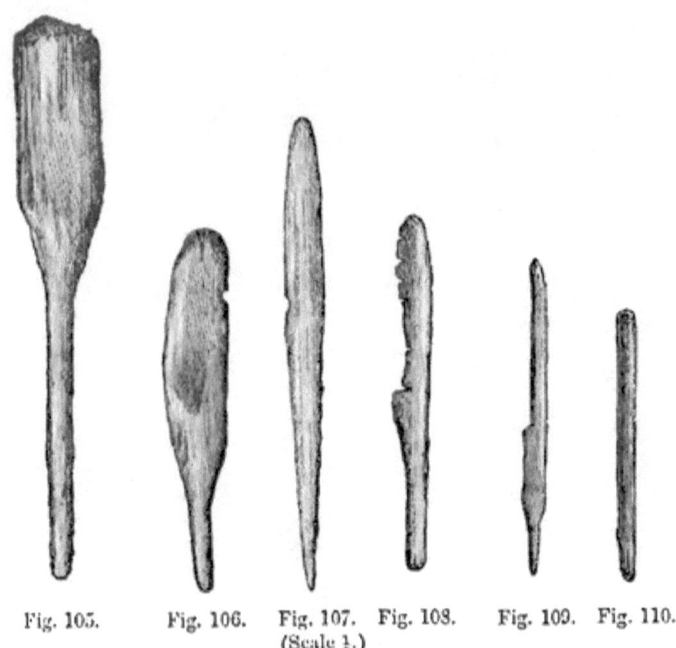

Fig. 105.   Fig. 106.   Fig. 107.   Fig. 108.   Fig. 109.   Fig. 110.
(Scale ¼.)

Figs. 111 to 115 represent the various kinds of pins which were abundantly met with all over the crannog.

Fig. 115 is 14 inches long, 2 broad, and 1⅛ thick; the hole in it measures 1⅝ by ¾ inch.

3. *Agricultural Implements, etc.*—Fig. 116. Mallet, head of which is 10 inches long and 16 in circumference; handle is 9 inches long and 5 in circumference.

Fig. 117.   Scraper or hoe, 10 inches long and 4 broad;

was cut out of a trunk of a tree, and had natural branch
formed into a handle.

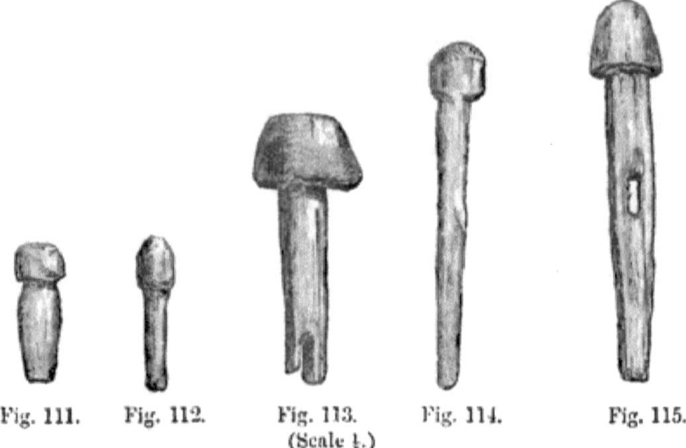

Fig. 111.   Fig. 112.   Fig. 113.   Fig. 114.   Fig. 115.
                              (Scale ⅓.)

Fig. 118.   Implement like boot or ploughshare, 10 inches
long and 12 round the middle.

Fig. 119.   Polished implement, 9 inches long, 5¼ broad,

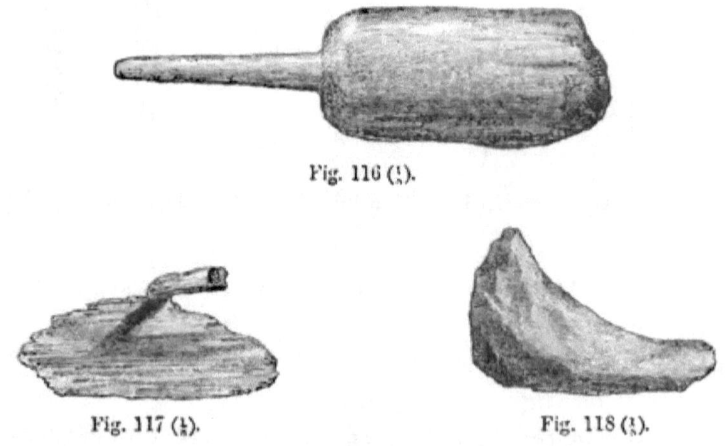

Fig. 116 (⅓).

Fig. 117 (½).   Fig. 118 (⅓).

and 2 thick (through the hole). The lower surface is flatter
than the upper, and slightly curved upwards longitudinally.

Fig. 120.   Horseshoe-shaped implement, 2 inches thick

and 2 deep at curve; greatest breadth $4\frac{1}{2}$ inches from the tips of the horns; depth of hollow $3\frac{1}{2}$ inches.

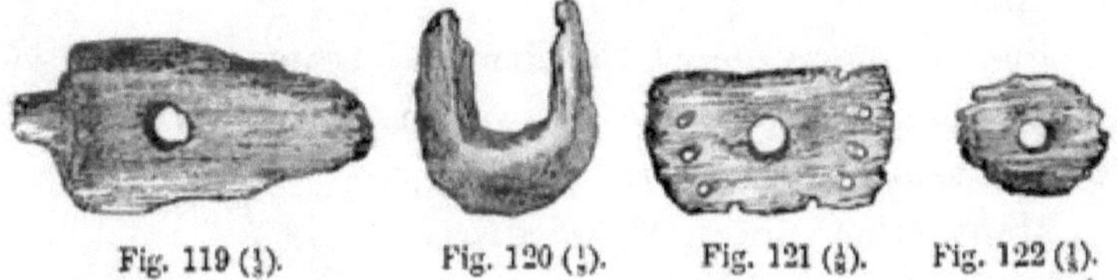

Fig. 119 ($\frac{1}{4}$).   Fig. 120 ($\frac{1}{4}$).   Fig. 121 ($\frac{1}{8}$).   Fig. 122 ($\frac{1}{4}$).

Fig. 121.   Portion of a circular implement, about 8 inches in diameter, and having a round hole in centre and ten small holes along the margin (if the circle were completed at same rate there would be fifteen holes in the series). The central hole was $1\frac{1}{4}$ inch in diameter, and had a tightly-fitting plug when found. The other holes were narrower in the middle, and large enough to admit of a common lead pencil to pass through. They also slanted slightly inwards, so that their axes, if prolonged, would meet at a common point about 6 inches from the central hole, and in the line of its axis.

Fig. 122.   Circular wheel, with hole in its centre and pointed teeth at circumference; diameter $3\frac{1}{4}$ inches, ditto of hole $\frac{3}{4}$ inch, thickness $\frac{1}{2}$ an inch.

Fig. 123 ($\frac{1}{24}$).   Fig. 124 ($\frac{1}{24}$).

Fig. 123.   Smooth piece of wood, 25 inches by 15, with square hole at top and two round ones at sides. Several other portions of boards, containing curious-shaped holes, were found.

Fig. 124. Piece of wood like the back of a seat in a canoe, 28 inches long by 9 broad. It has a raised bead round the margin.

Fig. 36 shows one of the mortised beams with portion of its upright taken from the outer trench at north-east corner.

Many other pieces of wood have been collected which illustrate various points of interest. One has a square hole showing marks of a gouge; another has a similar hole, but indicates that it was cut out by a straight-edged implement like a small hatchet; while a third, being part of the round tenon of a prepared beam splintered off, contains a number of small holes with wooden pins, showing how it had been mended.

4. *Canoes, Paddles, etc.*—At the commencement of our explorations, as already mentioned, a canoe, hollowed out of a single oak trunk, was found about 100 yards north of the crannog. Its depth in the moss was well ascertained, owing to the fact that, though lying at the bottom of one of the original drains, it presented no obstruction to the flow of water, and consequently was then undisturbed. During the recent drainage all the drains were made a foot deeper, and hence its discovery. It measures 10 feet long, 2 feet 6 inches broad (inside), and 1 foot 9 inches

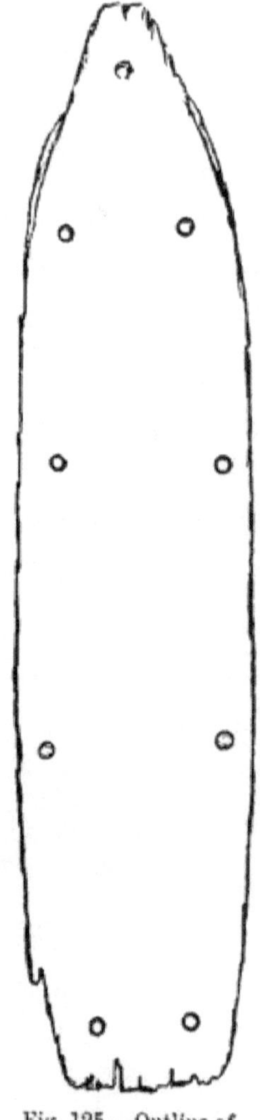

Fig. 125.—Outline of Canoe ($\frac{1}{24}$).

deep. The bottom is flat and 4 inches thick, but its sides are thin and rise up abruptly. There are nine holes in its bottom, arranged in two rows, and about 15 inches apart, with the odd one at the apex. These holes are perfectly round, and exactly 1 inch in diameter, and when the canoe was disinterred they were quite invisible, being all tightly plugged (Fig. 125).

The oak paddle here figured was found on the crannog. It is double-bladed, 4 feet 8 inches long and 5½ inches broad (Fig. 126).

Fig. 126.—Oak Paddle (¹⁄₁₄).

A large oar, together with the blade portion of another, was found on the margin of the crannog, which has already been described (see page 96).

When the original drainage was carried out some forty years ago, I understand that two canoes, each of which was about 12 feet long, were found in the bed of the lake on the south-west side of the crannog.

V. Objects of Metal.

(a.) *Articles made of Iron.*—1. A gouge, 8 inches long; stem 1¼ inch in circumference, slightly fluted before and behind; length of cutting edge ¾ of an inch; handle portion contained beautiful green crystals of vivianite (Fig. 127).

2. A chisel, length 10 inches; handle portion 3½ inches long; contains crystals and small remnant of bone handle; below handle there is a thick rim of iron; cutting edge measures only ½ an inch, and slopes equally on both sides. Top shows evidence of being hammered (Fig. 128).

3. Two knives. One (Fig. 129) has a blade 6 inches long, and a pointed portion for being inserted into a handle; found on a level with, and close to, the lowest hearth, along with fragments of its handle made of stag's horn. The other

(Fig. 47), found by a farmer in the débris long after it was thrown out of the trenches, was hafted on a different plan from the former, the end portion being broad and riveted to its handle by four iron rivets, which still remain. The blade is 6 inches long and much worn, being only ¼ to ½ inch in breadth, and the handle portion is 3½ inches long. Its position in the crannog is therefore uncertain.

Fig. 127.—Iron Gouge (¼).　Fig. 128.—Iron Chisel (¼).　Fig. 129.—Iron Knife (¼).

4. A small punch, 2½ inches long—locality uncertain (Fig. 130).

5. A bulky nail, some 4 inches long and ½ an inch in diameter, with large head; almost entirely converted into rust—locality uncertain.

6. A round pointed instrument, 11 inches long and 1¼ inch in circumference; its end portion is square, with sharp tip, as if adapted for insertion into a handle.

Fig. 130.
Iron Punch (¼).

7. An awl, 4 inches long.

8. Two spear-heads, 13 and $9\frac{1}{2}$ inches long, with sockets for wooden handles, portions of which still remain in both sockets. The larger of the two (Fig. 131) is prominently ribbed along its centre, and has a small copper rivet passing through the end of its socket. The other has only a very faint ridge along the centre of the blade (Fig. 132).

9. Five daggers. One (Fig. 133) has portion of a bone handle surrounded by a brass ferrule, and about an inch in front of this the remains of a guard are seen at the hilt of

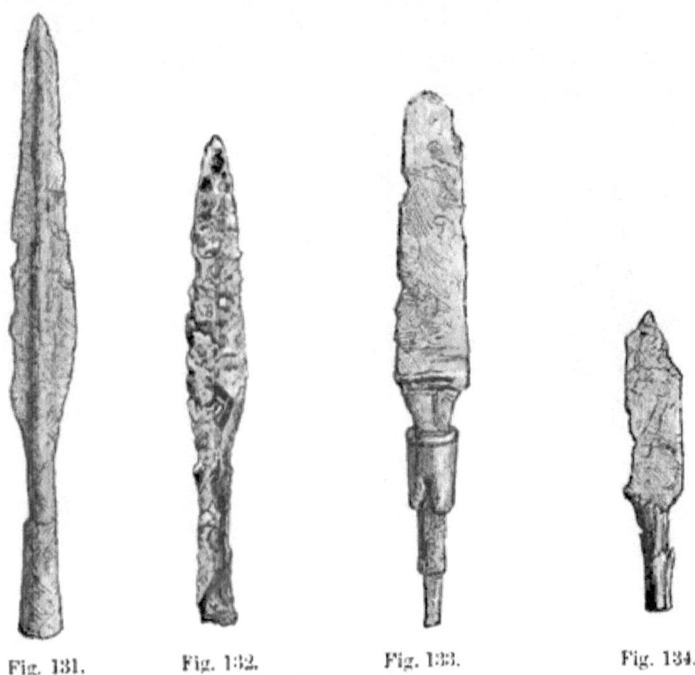

Fig. 131. Fig. 132. Fig. 133. Fig. 134.

Iron Weapons ($\frac{1}{4}$).

the blade; length of handle $3\frac{1}{4}$ inches, and circumference of ferrule $2\frac{1}{2}$ inches; the portion of the blade remaining is 6 inches long and rather more than an inch broad. Another, much corroded, has fragments of a wooden handle attached to it (Fig. 134). Fig. 135 represents a short pointed dagger, the blade of which is only $4\frac{1}{2}$ inches long, though at the hilt it is 1⅜ inch broad. The others are mere portions of the blades, one of which is drawn at Fig. 136.

10. A large ring. It is 3½ inches in diameter, and has a small portion of wood attached to one side (Fig. 137).

11. A saw, in three pieces, two of which were joined when found, and the third was lying a few feet apart. The length of the three portions together is 38 inches; average breadth is about 3 inches; teeth perfectly distinct and set. A small hole is seen at the end of one of the fragments. This relic was found at east side, external to the circle of stockades surrounding the log pavement (Fig. 43).

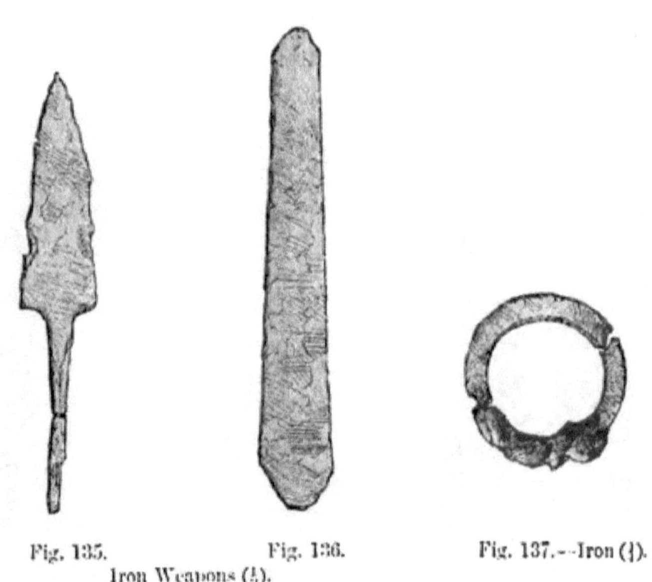

Fig. 135.   Fig. 136.   Fig. 137.—Iron (¼).
Iron Weapons (½).

12. An iron shears, 6¼ inches long (Fig. 138), was discovered on the site of the crannog by Captain William Gillon, 71st Highland Light Infantry, F.S.A. Scot. In presenting this shears, along with the hone-stone (Fig. 54), to the "National Collection," Captain Gillon thus describes the circumstances which led him to its discovery: "Having been in Ayrshire for the last five months, and within a short

distance of Lochlee farm, I have had several opportunities of visiting the site of the crannog, which was discovered there in 1878, and has since proved so rich in relics. . . . Although the crannog had been filled up, I determined to visit the site in the hope of finding some stray relic which might have escaped the eye of former explorers. In February 1881, Mr. Drummond, farmer, Pockenave, went with me, and as he had been present at the previous excavations, he showed me the most likely place for a 'find.'

"After looking about for half an hour, I was lucky enough to find the shears, which I forwarded to Professor Duns for examination, as I did not observe any articles similar to this in the Museum of Kilmarnock. On a subsequent occasion I found the hone-stone. I did not notice any hone-stones with a like groove in the Kilmarnock Museum, but on driving out to the crannog which is being excavated at Buston, near Kilmaurs, Dr. Munro showed me one which was of the same nature as the one I had found, only the groove was shorter and across the stone, and it had in addition a 'cupped hollow' in the centre, while this one has the groove lengthways." (See Fig. 54.)—*Proceed. Soc. Antiq. Scot.* vol. iii. new series, p. 247.[1]

13. Fig. 46 (p. 96) represents portion of a much corroded hatchet, about 6 inches long and 2 broad immediately below socket, but gets wider towards the cutting edge. Thickness through centre of socket is $1\frac{1}{2}$ inch. The back of socket was round, and had no projecting portion. Total weight $12\frac{1}{2}$ ounces. It had a small bit of the wooden handle in the socket when found.

---

[1] In the Museum of the Royal Irish Academy are several shears similar to the above, which were found on the crannogs of Dunshaughlin, Clonfinlough, and Strokestown; and in the York Museum there is also a collection of articles of the Early Iron Age and late Celtic period, amongst which I noticed five shears made on the same principle as that from Lochlee.

14. A curved portion of iron, like part of a door staple, found amongst débris, but locality undetermined.

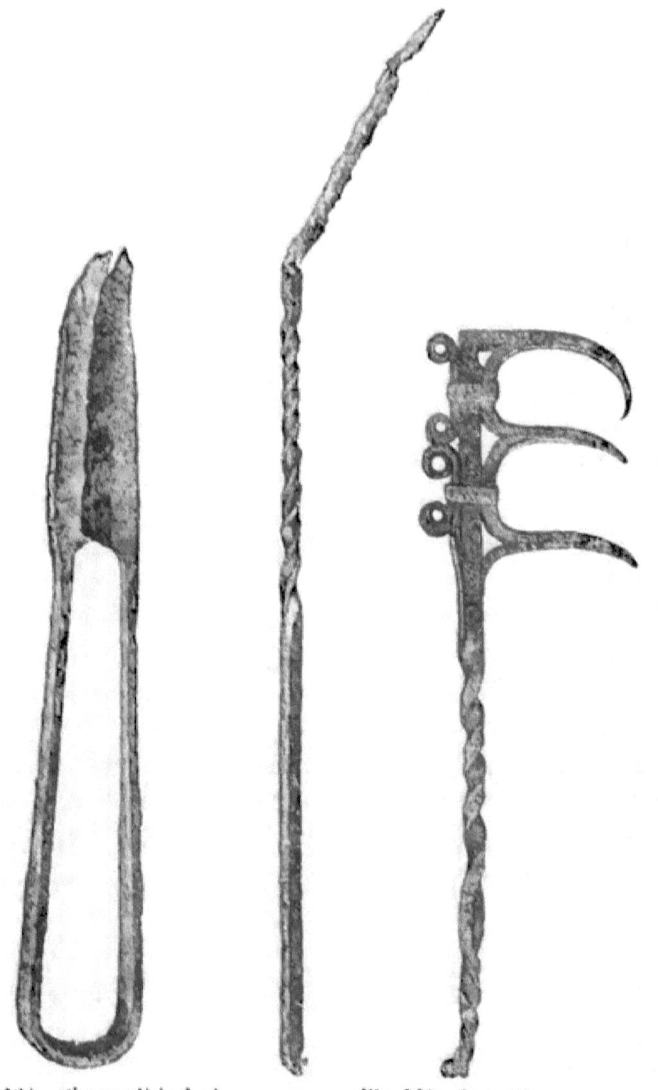

Fig. 138.   Shears (6¼ inches).   Fig. 139.—Iron (⅓).

15. A curious three-pronged implement (Fig. 139) was found, about 3 feet deep, in the large drain a few yards to

the south of crannog; the prongs are curved, very sharp at the points, and attached laterally; they are 2½ inches apart

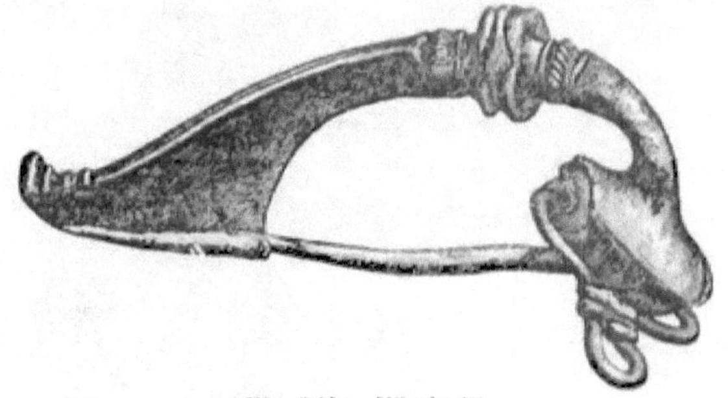

Fig. 140.—Fibula (¼).

and 4 inches long; a portion of the handle is twisted spirally; its total length is 3 feet 9 inches.

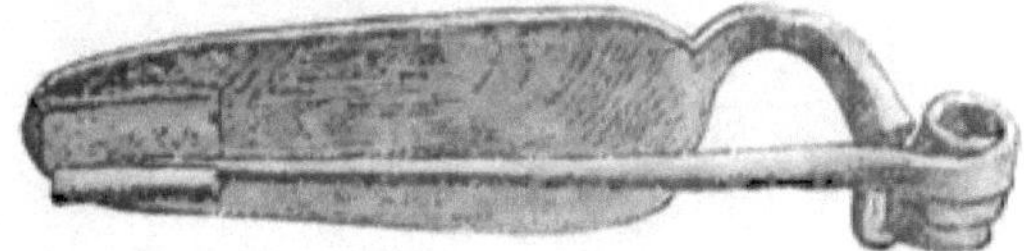

Fig. 141.—Fibula (¼).

16. A much corroded pickaxe was found about the middle of the lake area. The end of the axe portion is nearly 5 inches broad, and the whole length of the implement is 22 inches.

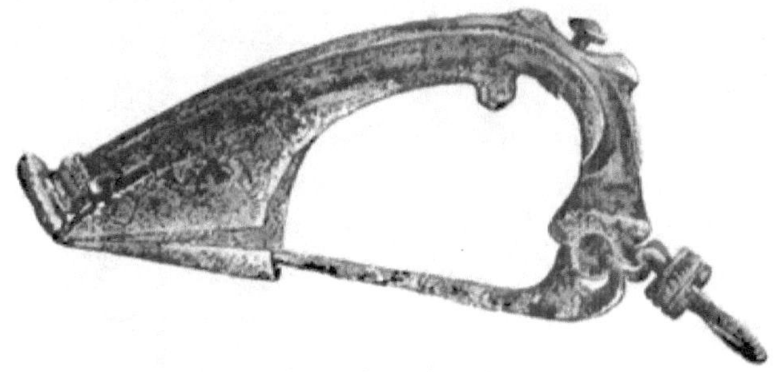

Fig. 142.—Fibula (¼).

(b.) *Articles made of Bronze or Brass.*—1. Two fibulæ, represented full size in Figs. 140 and 141, found about the centre of the refuse-heap. Figs. 142 and 143 represent side and back views of a third fibula, much more elaborately

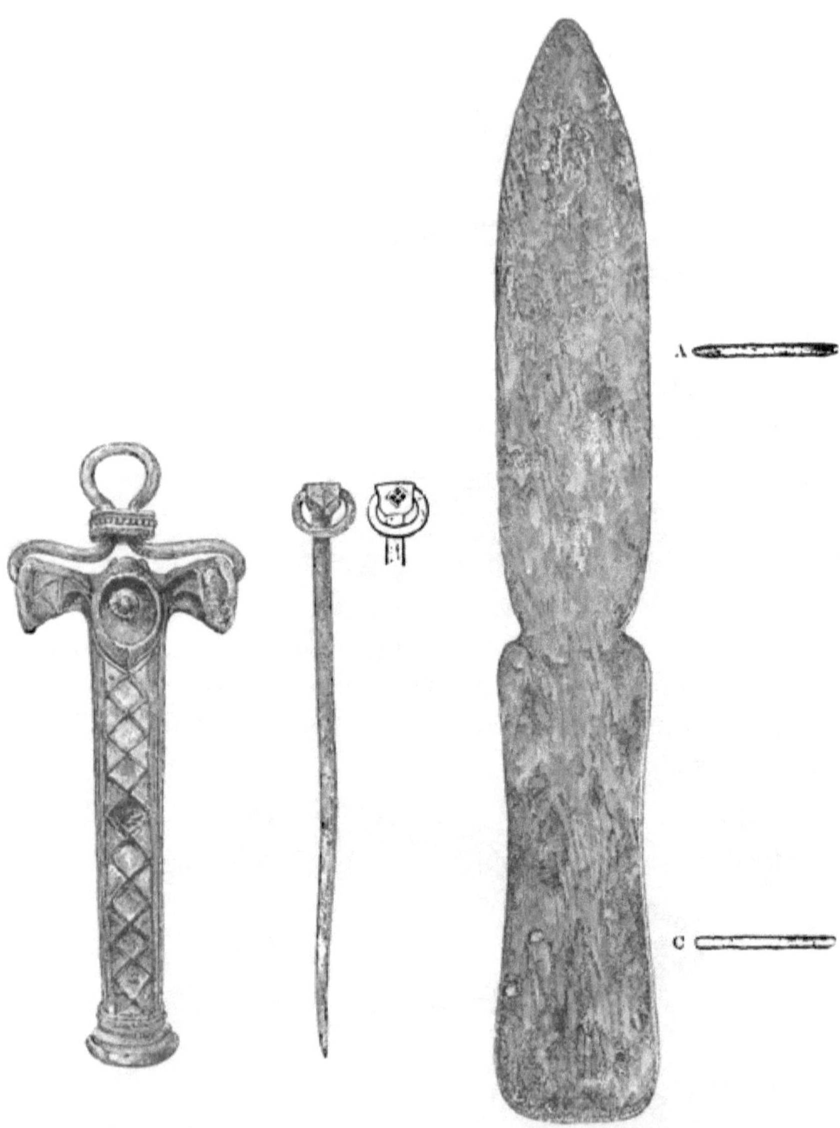

Fig. 143.—Back view
of Fibula, Fig. 142.

Fig. 144.—Bronze
Ring Pin (½).

Fig. 145.— Bronze Implement
with transverse sections (½).

ornamented, which was subsequently found in the débris

when closing up the trenches. The diamond-shaped spaces on its back were originally filled with enamel.

2. A bronze ring pin, 6 inches long. The square-shaped portion at the top has a different device on each side, and the shank from its middle to the point is ornamented on both sides (Fig. 144).[1]

3. A bronze spatula or dagger-shaped instrument. It is very well preserved, and although shaped like a dagger, the edges are not sharp. Its length is about 11¼ inches and breadth 1½ inch (Fig. 145).

---

[1] Colonel Gould Weston, F.S.A., has pointed out that one of these devices is a fylfot (croix gammée or swastika), an ancient symbol which in modern times has called forth a considerable amount of speculative writing. Its occurrence on four Irish monumental stones of the early Christian period has been the occasion of a recent article by the Bishop of Limerick (see *Proceed. of Royal Irish Acad.* vol. xxvii. part 3). The following extract from a paper, by M. Oscar Montelius, on the Sculptured Rocks of Sweden, is of interest as bearing on this point :—

" *La fréquence de la roue* ou du cercle crucifère (Fig. 11) et *l'absence totale de la croix gammée* (Fig. 12). Toutes deux sont, sans doute, des symboles religieux. La première (Fig. 11) qui se trouve très-souvent sur les monuments de l'âge du bronze, est presque totalement inconnue pendant l'âge du fer. La croix gammée (Fig. 12), au contraire, est très-fréquente pendant ce dernier âge ; je ne l'ai jamais vue sur les rochers sculptés dont nous parlons à présent."—*Compte-Rendu*, Congrès Inter. d'Anthrop. et d'Arch. Préhistorique, 7ᵐᵉ Session, 1874, Tom. i. pp. 459, 460.

Fig. 11.   Fig. 12.

See also Dr. Schliemann's works on the excavations at Troy and Mycenæ, where both these symbols are referred to as of frequent occurrence. In Dr. Schliemann's more recent work on Troy or " Ilios," an interesting account of the meaning and prevalence of this symbol among all nations is given.

It is found on some of the sculptured stones of Scotland. On a slab of greywacke from Craignarget, Gillespie, Glenluce, now deposited in the National Museum, Edinburgh, there is a cross on the upper part, with the sun and moon in the usual position above the arms, and two small crosses underneath, and below them a fylfot or swastika, together with cup-marks and concentric circles and various other devices. (See woodcut page 251, vol. iii. *Proceed. Soc. Antiq. Scot.*, new series.)

It also occurs on the famous Newton stone, along with two inscrip-

4. Portion of strong wire, 4 inches long, showing evidence of having been in the fire.

5. Thin spiral finger-ring (Fig. 146).

6. Fig. 147 represents a curious bronze object about $3\frac{3}{8}$ inches long; diameter of ring portion is 1 inch; the trans-

Fig. 146.—Finger Ring ($\frac{1}{1}$).   Fig. 147 ($\frac{1}{2}$).

verse bar at the other end is slit longitudinally and pierced transversely by a small hole about its centre.

7. Curved and slightly grooved bronze wire, $2\frac{1}{4}$ inches long, and precisely similar to the upper portion of a modern safety-pin.

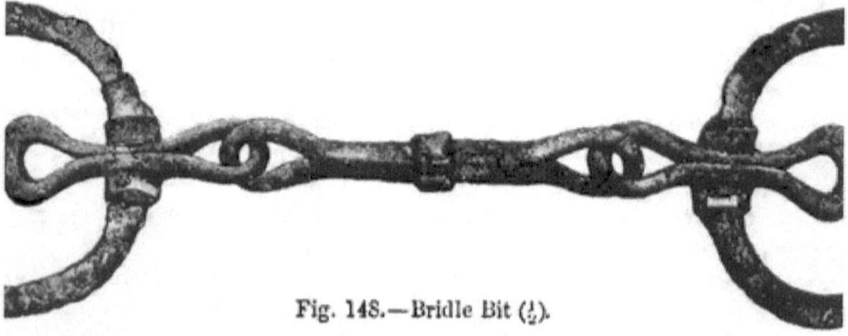

Fig. 148.—Bridle Bit ($\frac{1}{4}$).

(_e._) _Article made of Iron and Bronze._—A bridle bit. This consists of two large rings and a centre-piece. Its extreme length is $10\frac{1}{2}$ inches, the outer diameter of the rings is rather

tions, one of which is written in "Roman minuscular letters of an exceedingly debased form," and the other in Ogham characters, as well as on several other monuments of Christian time.

According to Dr. Joseph Anderson, although of Pagan origin, the fylfot has become a Christian symbol from the fourth to the fourteenth or fifteenth century.—(_Scotland in Early Christian Times_, vol. ii. p. 218.)

It is seen also in a mosaic pavement in the recently explored Roman villa in the Isle of Wight. (See Report by Cornelius Nicholson, F.G.S.)

less than 3 inches, and the centre-piece, which is entirely made of iron, is 3¾ inches long. The rings are partly iron and partly bronze, the circular portion being iron and the rest bronze. The bronze portion is 2½ inches long, and has two eyes or loops, one of which is attached to the centre-piece and the other free. This interesting relic was turned up by two visitors poking with a stick at the south-east corner of the refuse-heap (Fig. 148).

A round knob of lead, as if intended for the hilt of a hand weapon, was found very near the surface of the mound.

VI. Miscellaneous Objects.

1. *Carved Wood.*—Perhaps the most interesting of all the relics discovered on the crannog is a small piece of ash wood, about 5 inches square, having curious diagrams carved on both sides. On one side (Fig. 149), three equidistant spiral grooves, with corresponding ridges between, start from near a common centre and radiate outwards till they join, at uniform distances, a common circle which surrounds the diagram. On the other side (Fig. 150) is a similar diagram, with this difference, that between the points of commencement of the spiral grooves there is a space left which is occupied by a small circular groove surrounding the central depression or point. This figure is surmounted and over-lapped by two convoluted and symmetrical grooves meeting each other in an elevated arch, with a small depression in its centre. The relic was found on the west side of the crannog, about 4 feet deep, and near the line of the horizontal raised beams.

2. *Fringe-like Objects.*—Another object which has excited considerable curiosity is an apparatus made like a fringe by simply plaiting together at one end the long stems of a kind of moss (Fig. 151). Portions of similar articles were found in three different parts of the crannog, and all deeply buried.

The one figured here, and the most neatly formed, was found
in the relic-bed ; another about a couple of yards north of the

Fig. 149.—Carved Wood (⅓).

fireplace, and others at the south-west side, a little external
to the area of the log pavement.   In this latter place a large

quantity was found, but although the evidence of having
been plaited at one end was quite distinct, the stems of the

Fig. 150.—Carved Wood—other side of Fig. 149 (⅓).

moss were not prepared with the same care as in the one
figured overleaf, as the leaflets were still adhering to them.

The cue or pigtail, described at page 95, seems to have been formed of the same material as the so-called girdles or fringes.

Fig. 151 ($\frac{1}{3}$).—Made from stems of a moss (*Polytrichum commune*).

3. *Leather Objects.*—Fig. 152 is the representation of a fragment of a curious object, consisting of two portions of thick leather kept together by stout square-cut copper nails. These nails are broader at one end than the other, and pass completely through the layers of leather, after which they appear to be slightly riveted. The relic, as it stands, contains six nails, arranged in two rows, three in each row, and measures $2\frac{1}{2}$ by 2 inches, but the marks of additional nails are seen all round. Several portions of leather were collected from time to time. On the occasion of Mr. Joseph Anderson's visit to Lochlee, he found a shoe in the stuff just

thrown out of the bottom of the outer trench at the south side of the crannog. Other portions were picked up on the surface of the trestle-work, showing marks of having been neatly sewed. Also two stout thongs, one with a slit at the end through which the other thong passed and then formed a knot, together with a portion of coarse leather about the size of the palm of my hand, were found near the junction of the gangway with the crannog.

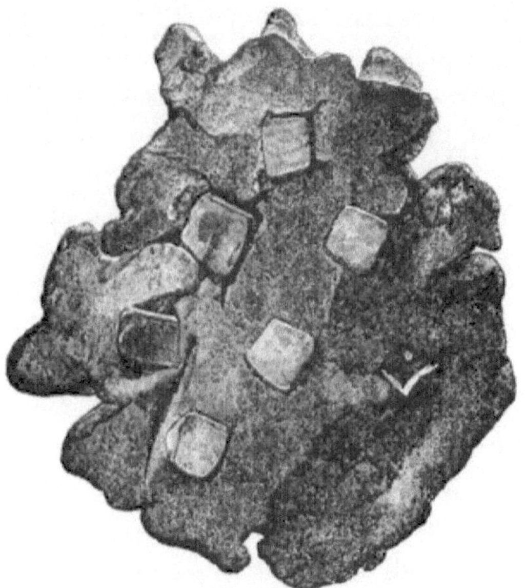

Fig. 152.—Portion of Leather with Copper Nails (⅓).

4. *Beads.*—Two fragments of beads, one fluted, the other smooth, and shaped like dumb-bells (Figs. 153 and 154).

 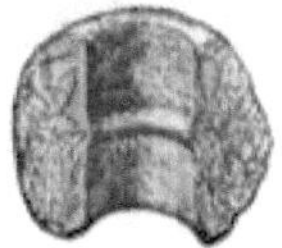 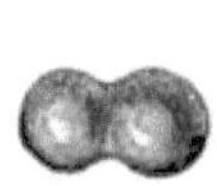 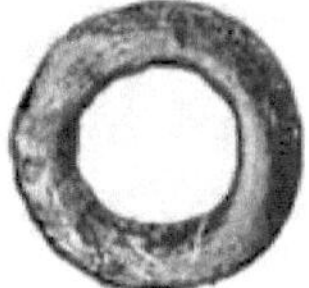

Fig. 153.—Beads (⅓).   Fig. 154.—Glass (⅓).   Fig. 155.—Bone Ring (⅓).

5. *Rings.*—A small bone (?) ring ¾ inch in diameter, and portion of another similar to the former, but a little larger.

6. *Pottery.*—(1.) The bottom of a dish made of reddish pottery, said to be Samian ware, was found in a drain close to the crannog.   Its diameter is 3½ inches.   (2.) Five small portions of a whitish unglazed ware, with circular striæ, as if made on the wheel, have been picked up in the débris after it had been wheeled out and lain exposed to the weather for some time, but the original situation of a single bit has not been determined.   These fragments might all belong to the same vessel, and two of them, though found at different parts and at different times, fit each other exactly.

7. *Crucible.*—A fragment of a small crucible coarsely made, and not unlike that found at Dowalton (Fig. 15), was for a long time mistaken for a bit of unglazed pottery.   It was found in the débris after it was wheeled out and had lain exposed for a while.

8. *Lignite, Jet, etc.*—(1.) A small bit of a black substance like a jet button.   (2.) Two portions of armlets made of lignite or jet, each about 2 inches long, were found near the wooden platform at the north-east corner.   One is a little thicker and coarser than the other, and forms part of a circle, which, if completed, would measure exactly 3 inches across (internal diameter).   The other is polished and of a jet black colour, internal diameter 2¾ inches.   A third fragment of a similar ornament was found in the débris when closing up the crannog.   It is more slender and has a

Fig. 156.   Fig. 157.

Portions of Jet Armlets (½).

smaller diameter than either of the others.   The portion of ring made of shale found at the bottom of the deep shaft below the log pavement is smaller than either of these, its internal diameter being only 2 inches, and its external 3¼ (Figs. 156 and 157).

9. *Tusks.*—The large number of boars' tusks met with, quite unconnected with the bones of the animal, especially in the relic-bed around the fireplaces, suggests the idea that they may have been used as implements. One only, however, was found to have decided marks of having been formed into a tool. It is $3\frac{1}{2}$ inches long, and very sharply pointed (Fig. 158).

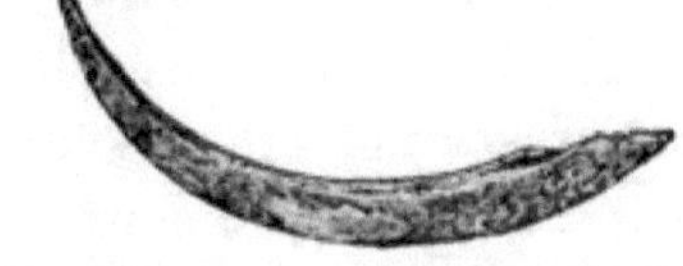

Fig. 158.—Implement of Boar's Tusk ($\frac{1}{2}$)

10. *Pigments.*—Several soft lumps of what appeared to be a blue and a red pigment were found, but they were not subjected to analysis. A specimen of the latter, which has been kept in a bottle, is now turning blue like the former

11. *Insect Cases.*—Large quantities of the horny coverings of insects like beetles were found in patches here and there, together with one or two brilliant-coloured *elytra*.

12. *Shell.*—One solitary shell was found near the fireplace, which I believe to be *Littorina littorea*.

THE FAUNA OF THE CRANNOG.

The following is the report of Professor Rolleston, M.D., F.R.S., F.S.A., on the fauna of the crannog:—

"Among the bones submitted to me by Robert Munro, M.D., Kilmarnock, and reported as having been procured from a crannog at Lochlee, the following animals have their skeletons represented :—

" The Ox, *Bos longifrons :* no proof of the presence of the wild variety.

" The Pig, *Sus scrofa*, variety *domestica*. I am not clear that the wild variety is represented here any more than in the specimens of the preceding species. (One fragment might have belonged to a wild individual, the molar No. 3 in it having all the pinnacles and eminences which have

given to the teeth of the *Suidæ*, as to the whole division of
non-ruminant Artiodactyles, the name Bunodont, worn away,
and having its grinding surface consequently reduced to one
single, however sinuous, continent of dentine bounded by
enamel.) As is well known,[1] the bones of an ill-tended and
ill-fed self-providing, so-called domestic pig, come to be very
like the bones of a thoroughly wild pig ; whilst, on the other
hand, it is also well known that very great variations exist
as to size within the limits even of the wild varieties of *Sus
scrofa.* But in the series now before me there is only one
fragment, consisting of the part of the lower jaw which
carries the last molar, and a part of the ascending ramus,
and of that last molar itself, which could, I think, by any
possibility be referred to the wild variety. And even here
such a reference could only be justified on the ground of the
great degradation which the cusps of the tooth have suffered,
it being usually the case that domestic pigs are not allowed
to live sufficiently long to get their teeth so worn down. I
have however to say that, both from this country and from
India, skulls of undoubtedly domestic animals of this species
have come into my hands, in which the teeth are worn down
far below the limits to which the molars of pigs are allowed
to be worn down by modern model-farm managers.

" The texture of the bone furnishes us with no indications,
its gloss and tenacity, if such it ever possessed, having been
entirely removed by its long maceration in water.

" It is however worth mentioning that this fragment from
a Scottish crannog exactly reproduces the contour of a fragment from the Starnberger See. (See Memoir on this "find "
in the *Archiv für Anthropologie*, viii. 1875.) In both, the

---

[1] See Nathusius, *Schweineschädel*, 1864, p. 147 ; Rütimeyer, *Basel.
Gesell. Naturforscher*, 1864, p. 161 ; Naumann, *Archiv für Anthrop.* viii.
p. 23, 1875 ; Studer, *Zürich. Mittheilung., Pfahlbauten*, 1876, xix. 3,
p. 67.

angle of the jaw has been knocked away, for the sake, doubt-
less, of the soft and succulent, and I may add sensitive,
substances it protected during life, and in both the posterior
molar has been left *in situ*, though much worn down. The
posterior molar, however, of the foreign specimen has that
superior development of its third molar, which, if Nathusius
(*Schweineschädel*, p. 49) had not taught us better, might have
been referred to domestication instead of to better food or
sexual (male) character. I owe this specimen to the kindness of J. E. Lee, Esq., F.G.S., and though I hesitate in the
case of the Scottish specimen, I have no hesitation in referring this one to the wild variety, as indeed it is referred
under the title of *Sus scrofa ferox* on the label it carried when
it came into my hands.

"The specimens of pigs' bones and of pigs' teeth are
numerous, but none other either of the bones or of the teeth
are of the size, strength, or proportions which would have
enabled their owners to hold their own as wild animals in
a country in which the wolf may still have existed.[1]

"The sheep, old dun-faced breed, *Ovis aries*, variety
brachyura. One nearly perfect skull of a sheep of the
variety which is known as *brachyura*,[2] from having a short
tail, but which also has the horns of the goat, set on, it is true,
with their long axis at a different angle from that which they
have in the true goat, but still in themselves of very much
the same shape. One lower jaw in this series has the concave posterior boundary, and the sinuosity anterior to its
angle, which goats usually and sheep only sometimes,
possess. It belonged, however, to an immature individual,
the posterior molar not having been evolved, and it cannot

---

[1] For reference to the bibliography of prehistoric Swine, see *Linnean
Soc. Trans.*, ser. ii., Zool. vol. i. 1877, p. 272.

[2] For reference to the history of this variety of sheep, see *British
Barrows*, p. 740.

be considered to positively prove the presence here of *Capra hircus.*

" The Red Deer (*Cervus elaphus*) is very abundantly represented in this series, especially by fragments of horns, some of which bear marks of having themselves been cut and sawn by other implements, whilst one or two may possibly have been used, as the tines of red deer so often were by the early British flint miners, as borers.

" The Roe Deer (*Cervus capreolus*) is only scantily, though unambiguously, represented in the collection from Lochlee.

" The Horse (*Equus caballus*) is represented by but a single shoulder-blade; it is of small dimensions relatively to most or all domestic breeds with which I am acquainted; this applies, however, to all the domestic animal remains found here.

" Reindeer (*Cervus tarandus*).—There are two more or less fragmentary portions of horns, which, after a good deal of comparison with other reindeer horns, and with fragments of red deer horns, I incline to set down as indicating the presence of the former animal in this collection. It is easy to separate reindeer horns from red deer horns when you have the entire antler before you, or even when you have the brow antler only, in most cases ; and it is usually easy to separate even a fragment if the fragment is fresh, because the surfaces of the horns in these two horns are different. But here the two fragmentary horns in question have no brow antler left, and their surfaces have been macerated so long as to have desquamated, or, to change from a medical to a geological metaphor, have been denuded a good deal. Still one fragment is, I think, too tabular, and the other is too tabular also, and that just below the origin of what in the red deer is known as the sur-royal antler, to be anything but a reindeer's.

" Writing for Scottish readers, I need not refer to Dr. J.

A. Smith's paper 'On Remains of the Reindeer in Scotland,' read before the Society of Antiquaries of Scotland, June 14, 1869, vol. viii. pt. i. pp. 186-223, nor to his references in that exhaustive memoir to preceding writers. But I may mention an additional reference which Dr. J. A. Smith, not being gifted with as much second sight as he is with insight, could not have then referred to, as it is contained in a book of more recent date than is his paper. This reference will be found in Mr. Joseph Anderson's edition of the *Orkneyinga Saga*, chap. vi. p. 182."

Regarding a subsequent consignment of bones and horns sent to Professor Rolleston, he writes as follows :—

"The only remark which I feel called upon to make relates to the bones and the teeth of the pig; the marrow cavity in the lower jaw of one of the pigs, a young specimen, containing a large quantity of crystals, and the teeth of the older pigs showing a great deal of wear for the teeth of what were, I think, domesticated swine. The crystals were analysed by W. W. Fisher, Esq., of the Chemical Department in the Oxford Museum, and found to be vivianite as supposed. It is not uncommon to have bones from pre-historic 'finds,' which have been much acted on either by fire or water, thus coloured by double decomposition of the bone phosphate with some iron salt furnished either from the bone and flesh or otherwise.

"The horns" (all the worked ones in the collection) "received a few days ago are all of Red Deer (*Cervus elaphus*), except one, which is of *Cervus capreolus*. With this consignment came one bone, or rather the ulna and radius of a *Bos longifrons*, more or less fused into one bone. The horn of the Roe is rather a large one."

THE FLORA OF THE CRANNOG.

As there appears to be some difference of opinion among botanists as to whether certain trees, now common in our forests, such as elm and beech, are indigenous to Scotland, my attention was directed at an early stage of the investigations at Lochlee to the importance of determining the different kinds of wood used in the structure of the crannog. Accordingly, I collected specimens of the wood and other vegetal remains encountered during the excavations, and in due time forwarded them to Professor Balfour, Edinburgh, who had kindly agreed to examine and report upon them, but unfortunately, owing to ill-health, he was unable to do so, and the box containing the specimens, after lying in Edinburgh for some weeks, was returned unopened. Ultimately, however, Dr. Bayley Balfour, Professor of Botany in the University of Glasgow, undertook this task, and it is to him I am therefore indebted for the following report :—

"I shall send by train to-morrow the box of Lochlee vegetable remains. I have examined them carefully, and you will find each specimen numbered, the numbers corresponding with those in the appended list. There is not so much variety in the wood as I anticipated, and I am surprised to find neither oak nor fir. The tissue of the wood is in most cases considerably decomposed, the wood cells, as might be expected, being most affected. Betwixt alder (*Alnus glutinosa*, L.) and poplar (*Populus tremula*, L.), the only indigenous species, there is really very little difference in wood structure, and indeed birch (*Betula alba*, L.) and hazel (*Corylus Avellana*, L.) are not far removed, so that that when the texture of the wood is much compressed, and decomposition has progressed, an identification is somewhat hazardous, and I have therefore queried my determination in some cases. No beech occurs amongst the specimens you sent me."

The following is a summary of the detailed list :—

I. *Brushwood, etc.*—The various specimens of wood which were selected from below the log pavement have been classified as belonging to one or other of the following trees, viz., birch (*Betula alba*), hazel (*Corylus Avellana*), alder (*Alnus glutinosa*), and willow (*Salix*, sp.).

II. *Wooden Relics.*—One of the implements (Fig. 118), which appeared to be made of a different kind of wood from any of the rest, has been identified by Dr. Balfour as elm (*Ulmus montana*, Sm.); and the piece of board with the carved diagrams (see pages 134, 135) is found to be ash (*Fraxinus excelsior*, L.). The rest of the relics were not submitted to Dr. Balfour, as they had so crumbled into dust (except those made of oak, all of which were easily recognised) that their identification appeared impossible.

III. Among the remaining vegetal remains collected from the débris above the log pavement, Dr. Balfour has identified the following species :—

"(1.) *Hypnum (Hylocomium) splendens*, Dill. This specimen I submitted for confirmation to Mr. Hobkirk of Huddersfield, and after the most careful examination he refers it to the above.

"(2.) *Dædalea quercina*, P. This I submitted to Dr. M. C. Cooke for confirmation, and he remarks, 'Must be a thin form of that species; but of course it is very much discoloured, and hence difficult to determine.'

"(3.) *Bovista nigrescens*, P.

"(4.) *Polyporus igniarius*, Fr. This and the preceding are Dr. M. C. Cooke's identification.

"(5.) *Polytrichum commune*, L. (Portions from the fringe-like girdles (Fig. 151) and the pigtail-like object described at page 95 were thus labelled.)

"(6.) *Pteris aquilina*, L.

"(7.) Several masses containing roots and root leaves of a monocotyledonous plant with equitant leaves, heather stems, and rhizomes of fern.

"(8.) Portions of birch bark in stripes rolled together like a ball of thread.

"(9.) Hazel nuts. One gnawed by a *squirrel?* If, as I conjecture, it has been done by a squirrel, it is interesting, as affording evidence of their occurrence in this locality.

"I am sorry I am unable to be more definite in many cases. The masses made up of monocotyledonous plants would not repay a more extended examination."

General Remarks.

To extract from the above investigations, however suggestive the results may appear, a life history, as it were, of the crannog, or indeed much reliable information regarding the habits of the Celtic races who flourished in the neighbourhood during the period of its existence, would be presumptuous on my part, if not beyond the scope of legitimate inference, especially in face of the meagre results hitherto obtained from Scottish crannogs. The completeness with which the operations have been executed, together with the great variety of relics found, cannot fail to make the Lochlee crannog a standard of comparison for future discoveries of a similar character, at least for some time to come, and hence it was essential to have the present report free from all speculative opinions. I have therefore up to this point entirely confined myself to matters of fact which have come under my own direct cognisance; and as for the relics, I have simply endeavoured to describe them accurately, leaving it to experienced archæologists to determine their historical value. There are however a few points, bearing on the antiquity and duration of the crannog, which, though undoubtedly included in the category of the speculative, I wish to state, as they could only be made by one conversant with all the phases of the excavations; but which, after this caution, must be taken *cum grano salis.*

1. *Position of Relics.*—As many of the relics, if judged independently of the rest and their surroundings, would be taken as good representatives of the three so-called ages of Stone, Bronze, and Iron, it is but natural for the reader to inquire if superposition has defined them by a corresponding relationship. On this point I offer no dubious opinion. The polished stone celt, Fig. 55, and the knife, Fig. 129, were found almost in juxtaposition about the level of the lowest fire-place. Though the hammer-stones, as a rule, were more abundant in the lower strata, yet the very first thing indicating human art which was found, when we commenced to dig towards the centre of the mound, was a hammer-stone. Almost all the horn implements were found at or below the level of the first-discovered pavement, and three-fourths of the querns were found above it. Below the same level, and around the hearths, tusks of boars were numerous, whereas almost none were found above it; and in the midden pigs' jaws and teeth were found only at its lowest stratum. Various inferences might be drawn from these remarks, which my readers can do for themselves.

2. *Character of the Wood-work.*—From the discovery, in the deep section made below the log pavement, of beams with tenons and mortised holes, and large trunks having their branches lopped off as if with a hatchet, it would appear that the first constructors of the crannog were well acquainted with the use of metal tools. It may also be noted that the base of the wood-work of the island was at least 14 feet (after making allowance for the extra height of the mound) below the surface of the field, whereas that of

the gangway, within a few feet of the crannog, was only 10 feet. This difference of 4 feet could hardly be accounted for by the inequality of the bed of the lake, as the field appeared here to be quite level, so that we must infer either that it was due to the absolute weight of the island on the portion of the bed of the lake directly below it, or that the gangway was of more recent construction than the island. After all these doubtful elements are eliminated, there remains the important fact that since the island was constructed, 10 feet of silt, at least, had accumulated around it.

That broken planks and old mortised beams were mixed up with the trestle-work in various places, would support the idea of a prior structure; while evidence that the whole superstructure had at one or more times been destroyed by fire was quite conclusive. According to this theory, the elaborate mortised beams at the north-east corner might have been a landing-stage, but, in their present position, they are quite inexplicable.

3. *Level of Lake.*—Amongst the problems of a discursive character here referred to, perhaps there is none of greater interest than that which deals with the cause and effect of the change that has taken place in the level of the lake. From eye-witnesses we know that, before the first drainage was carried out, the mound used to be covered with water in the winter-time; and Mr. Charles Reid tells me that the line of level which he has adopted in measuring for the plan of the lake is 8 feet 7 inches above the log pavement. Now the area assigned to the lake by Mr. Reid is considerably less than what the old residenters of the district make it out to have been, as several of them have stated that they had seen its waters extend beyond the road on the west side (see Plan), and yet from his data the depth of water would just cover the highest part of the mound, which, it will be remembered, was about the same height above the log

pavement. Originally the island must have been higher than the lake, but allowing that the log pavement was only 3 feet above the surface of the water, we have at least 11 feet of change of level to account for. This phenomenon could only be caused by a sinking of the whole island, or a rising of the water, or a combination of both causes. I do not think that the weight of the island and its superincumbent mass would press so heavily on the bottom of the lake as to cause it to sink much, since the enormous amount of wood-work, of which it consisted, being lighter than water, would have a corresponding buoyant effect, and so help to counter-act the weight of the aërial portion. Nor has any great compression of its substance taken place from decay, because in the course of making the deep section under the log pave-ment, we found the contour of the large trees quite sym-metrical and perfectly round; and although the wood was very soft it was not compressed, owing to its being com-pletely saturated with water, which of course is virtually incompressible. Although I have often seen small brush-wood flattened from pressure, yet I have never seen this effect produced on a branch larger than my wrist, and only in one instance did I notice it on a piece of wood of this size. Moreover, the gangway, which certainly could not sink from its weight, was deeply buried, its uppermost horizontal beams being not less than 7 feet below the surface of the field. We must therefore fall back on the only other alter-natives, and assign this change in the relative position of the crannog and level of the lake either to a general compres-sion and sinking of the bed of the lake or to the rising of the water. The latter result is somewhat unusual, because running water, having a tendency to deepen its channel, and the accumulation of sedimentary deposits, often produce an opposite effect, and cause the complete drainage of lake basins. I have therefore carefully examined the outlet of the lake to ascertain if possible the causes that led to this rise in its bed.

Its natural outflow was at the south-east corner, and the little stream, after running southwards for a few yards, quickly turned westwards into a narrow valley which wended towards Fail Loch. Just at this abrupt turning the background rose somewhat steeply to the south, so that the termination of the valley as it entered into the Lochlee basin was very liable to be obstructed by débris washed from the slopes above. Besides being thus favourably situated to catch washed-down materials, it is probable that during the summer the surplus water would be very scanty, and vegetation abundant, so that in the course of time the bed of the outlet would gradually be raised. A section cut across the outlet would readily disclose the sequence of the silted materials, had it not been that the soil was disturbed by a deep covered drain, which was made when the first drainage operations were executed, and ran along its whole course. Also, I understand that previous to this the lake was used as a mill-dam. We cannot therefore get rid of the elements of uncertainty in any calculations which might be based on the change of the level of the lake and the accumulation of silt in its bed and outlet.

I may however mention, on the grounds already stated, that since the foundations of the crannog were laid, the increase in the bed of the lake in its vicinity cannot have been less than 10 or 14 feet; and 11 feet is the lowest estimate that I can assign to the submergence of its surface. On the supposition that this apparent encroachment of the water was uniform, and since the last fireplace was about 6 feet above the lowest, and allowing 1½ feet for the time the former was used, we have then the total period of occupancy of the crannog represented by 7½ feet of total submergence. We have no means of comparing this period

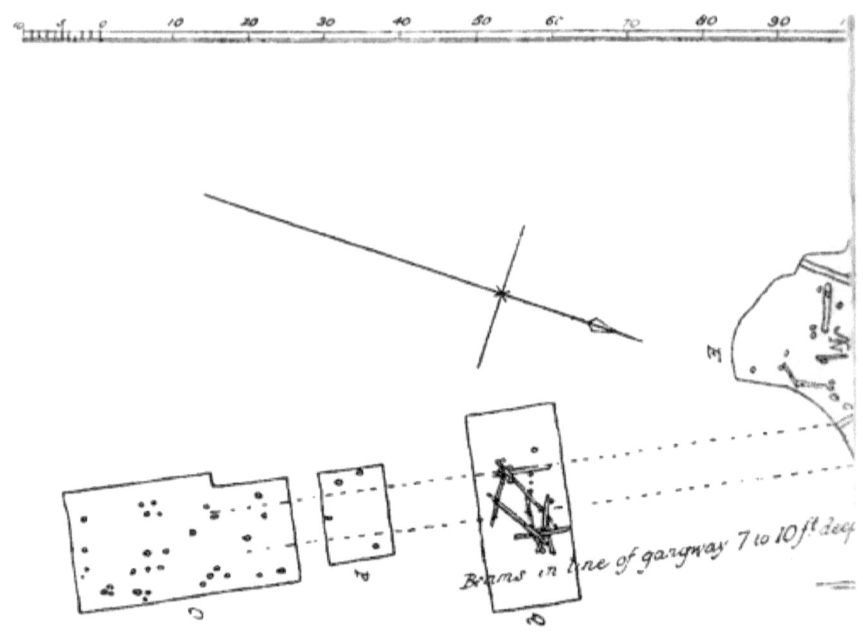

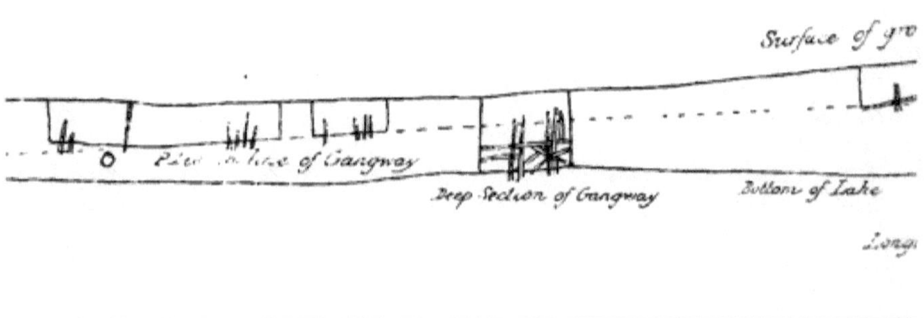

PLAN & SECTIONS o

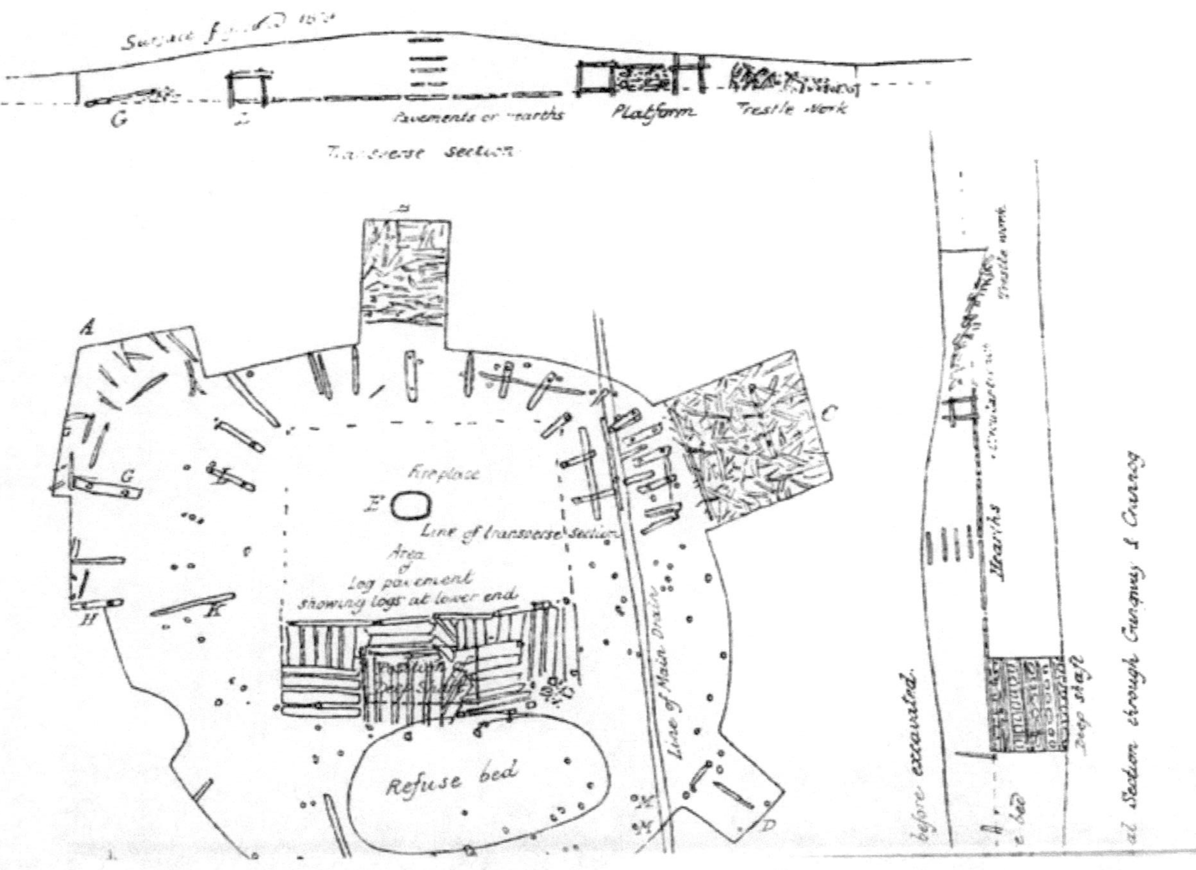

CRANNOG AT LOCHLEE.

with its representation in so many feet of lake sediment, but I may state that since the canoe, found about 100 yards from the crannog, was abandoned, no less than 5 feet of this mossy lake sediment accumulated over it.

The composition of the silt forming the bed of the lake, especially near the crannog, as already described at pp. 98, 99, points to the fact that for centuries the increase was due principally to the decomposition of vegetable matters, while latterly it was caused more by a deposition of fine clay; and when excavating along the line of the gangway we had an opportunity of verifying the regularity of this succession. A change so marked in the sediment can only be accounted for by a corresponding change in the surrounding scenery, and no explanation is more likely than that the primeval forests had given place to the inroads of agriculture, when some of the upturned virgin soil would be washed down, as it still is, by every trickling rill that finds its way into this lake basin.

[The selection of bones sent to Professor Rolleston for examination is now deposited in the Anatomical Museum at Oxford, and all the rest of the relics are located in the Museum attached to the Burns Monument at Kilmarnock.]

CHAPTER IV.

SUBSEQUENT RESEARCHES AND DISCOVERIES AT FRIARS' CARSE,
LOCHSPOUTS, BARHAPPLE, AND BUSTON.

BEFORE the interesting series of objects obtained from the excavation of the Lochlee crannog, and described in the previous chapter, could be properly illustrated and efficiently brought under the notice of antiquaries, other promising "finds" of a similar nature were announced from various quarters. All these have now, as far as practicable, been carefully investigated, with results, in some respects, even more remarkable than the former. To a description of these further excavations and discoveries I propose to devote this chapter, commencing with that of a lake-dwelling at Friars' Carse, in Dumfriesshire, and following it up by others in the order of their discovery.

SECTION I.

Notes of a Crannog at Friars' Carse, Dumfriesshire.

Early in the summer of 1879 I was informed that, during the autumn of the preceding year, a lake-dwelling had been exposed in a small loch at Friars' Carse, Dumfriesshire, and being then engaged in drawing up a report of the excavations made at Lochlee, I was anxious to have an opportunity of comparing the results obtained from the former with those of the latter. This opportunity was afforded me by the Rev. Mr. Landsborough, who, being also interested in such

discoveries, made arrangements with his friend, Dr. Grierson
of Thornhill, to conduct us to Friars' Carse.   After inspecting a canoe, some fragments of pottery, and a few other
things from this crannog, then deposited in Dr. Grierson's
museum, we drove off to inspect the structure itself.   Its
site was a small pear-shaped basin situated behind a wooded
knoll, close to the Parliamentary road to Dumfries, and in
the midst of a well-cultivated but singularly undulating
district.   By deepening the outlet of this lake to the extent
of 2 feet, a partial drainage was effected, which reduced its
area from 10 to 3 acres.   It was then that it became generally known that a small bushy island near the middle of
the loch had been artificially constructed of oak logs and
trunks of trees.   As the weather was dry for some weeks
previous to our visit, and the water particularly low, we
readily stepped on to the island, over what appeared to have
been the old bed of the lake, then presenting a hard, crisp,
and dried-up surface of aquatic plants.   The island was
nearly circular in shape, strongly built, and surrounded by
piles, some of which, however, were only visible through the
water.   The log pavement, which by this time had been
completely bared, was composed of parallel beams of oak,
apparently arranged in groups, lying in various directions,
and firmly united together by the overlapping and sometimes
mortising of their ends.   Its level was from 1 to 2 feet above
that of the water, but at the margin of the island there was
a large quantity of stones, especially on its north side, *i.e.*
the side towards the deepest portion of the lake, and most
distant from the outlet.   Through these stones, which shelved
under the water, a few heads of the surrounding piles projected, some of which were also visible above the water.
Some mortised holes were here and there to be seen in the
horizontal beams, but there was no trace of a raised breastwork surrounding the wooden pavement—thus differing in

this respect from the crannog at Lochlee, and agreeing with that at Lochspouts. In the centre were a few ends of uprights, in rectangular rows, seemingly the remains of partitions, one of which I traced for 40 feet in a straight line.

Upon inquiring where the rubbish removed from the island was located, we were informed that it had been wheeled to the west side of the crannog, and heaped up just close to where we had stepped across to the island. Here it lay for some days; but one morning, to the great astonishment of the workmen, it was found to have entirely disappeared. Upon examination, it turned out that the apparently dry land was a matted crust of mud and roots of aquatic plants, which virtually floated over the water, and suddenly gave way under the accumulated weight, and so buried the whole mass in the water beneath. With this singular and unfortunate catastrophe terminated all further prospects of finding relics.

My examination of the crannog was then of a very limited character, and hence, when I came to require more definite information, I found it necessary to revisit the locality. This visit took place so recently as the 31st January 1882, and, although a day by no means suitable for such investigations, I am glad to say that through the courtesy and kindness of the proprietor, Thomas Nelson, Esq., who was personally conversant with the drainage operations, and took much interest in the Lake-Dwelling, the following additional details were procured :—

The island is slightly oval in shape, and, including the partially submerged zone in which the piles were noticed, measures 80 by 70 feet. Near its centre the débris was from 2 to 3 feet thick, and formed a sort of mound containing ashes, charcoal, and some bones. Here the fragments of pottery afterwards described were found.

A circular portion of the log pavement, near the centre,

was covered with small stones as if to protect it from fire; also some remains of clay flooring were observed in other parts of the island.

Regarding the deeper structures little can be said. Mr. Nelson attempted to cut a hole through the timber, and, as far as the water allowed the men to penetrate, he saw nothing but layer upon layer of oak beams lying transversely upon each other. Judging, however, from the solidity and firmness of the island, the great size of some of the logs, and the depth of the loch (still about 12 feet a little to the west of the island), the total thickness of this immense mass of timber cannot, I should say, be less than 12 or 16 feet.

Mr. Nelson has directed my attention to the following notice of this island in the *Antiquities of Scotland*, by Grose, vol. i. p. 146.

FRIARS' CARSE, IN NITHSDALE.

" Here was a cell dependent on the rich abbey of Melrose, which, at the Reformation, was granted by the Commendator to the Laird of Elliesland, a cadet of the Kirkpatricks of Closeburne. From whom it passed to the Maxwells of Tinwald, and from them to the Barncleugh family, also cadets of the Lords of Maxwell. From whom it went to the Riddells, of Glenriddell, the present possessors. The old refectory, or dining-room, had walls 8 feet thick, and the chimney was 12 feet wide. This old building having become ruinous, was pulled down in 1773, to make way for the present house.

" Near the house was the Lough, which was the fish-pond of the friary. In the middle of which is a very curious artificial island, founded upon large piles and planks of oak, where the monks lodged their valuable effects when the English made an inroad into Strathnith."

From the above quotation it would appear that this

structure has not ceased to be an island by becoming submerged, like most of the other lake-dwellings hitherto noticed. The surface of the log pavement is at present about 18 inches above the water-level, so that, before the recent drainage, it would be 6 inches below it, but, originally, it must have been 3 or 4 feet above the ordinary level of the loch. Hence, on the supposition that no great alteration was made on the lake area by former cuttings, the maximum amount of subsidence would not be more than 5 feet.

Canoe.—About 60 yards from the island, while making the cut for drainage, a canoe was found, "deeply imbedded" in the mud (about 4 feet). It now lies in Dr. Grierson's museum at Thornhill, but it has become so shrivelled and distorted that it would be difficult to recognise it as a dug-out canoe. From Dr. Grierson's description of it,

Fig. 159.—Perforated Axe-hammer Head (⅓).

shortly after discovery, it appears to have been 22 feet long and 2 feet 10 inches broad. The prow was the root end of the tree, and tapered to a point, but the stern, which was squarely cut, was closed by a flat stern-piece fitting into a groove.

A neatly formed paddle was found on the west side of the loch. Its length is 3 feet 10 inches, of which the blade takes up 1 foot 6 inches by 5 inches broad.

The ponderous axe hammer-head here figured (Fig. 159), was found on the west side of the loch along with the paddle. "It was about 2 feet below the present surface, and about 30 yards from the island, at a place where the ground was firmer and might have been a landing-place from the island." It is made of hard whinstone, and measures 10 inches in length, 5 inches in breadth, and a shade less than 3 inches in depth. It is perforated by a round shaft-hole, 2 inches in diameter, but tapers slightly from both surfaces to the middle.

Pottery.—Two handles of jars with a yellowish glaze, inclining in some parts to a green and in others to a reddish-brown colour.

Fig. 160.—Pottery (⅔).

Fig. 161.—Pottery (⅔).

Two fragments of a large dish were ornamented by a series of little pits regularly grouped together in the form of bands as represented in Figs. 160 and 161. These bands were formed of three parallel rows of pits, which in the larger fragment radiated from the base upwards at a distance of about an inch, but became a little wider at the bulge of the dish. The other fragment (Fig. 161) is too small to indicate the direction of the pitted rows, but

the band is decidedly raised above the general surface of
the vessel—a feature which is only partially noticeable
in the former, just at the upper termination of the rows.
The little indentations are irregularly shaped, but, from
a repetition of the same peculiarities in their form in
each row, it is clear that they were made by a small
trident-like implement.

All these fragments of pottery were made of a fine
reddish clay, mixed with coarse sand or small quartz pebbles.

The only other relics were a circular stone polisher,
$2\frac{1}{2}$ inches in diameter, and an oval-shaped mass of vitreous
paste.

SECTION II.

*Notice of the Excavation of a Crannog at Lochspouts,
near Kilkerran.*

Situation of Crannog.—Lochspouts is situated about
three miles to the south-west of Maybole, in the parish
of Kirkoswald, and on the property of the Right Honourable
Sir James Fergusson, of Kilkerran, Bart., K.C.M.G., LL.D.
It is a small lake basin, somewhat oval in shape, and
ensconced at the base of hilly ground, which encompasses
it, except towards the north, where a narrow trap-dike runs
across and cuts it off from the open valley beyond. It is
thus a natural dam, formed in the face of a declivity, which,
beyond the trap ridge, still continues to slope rapidly downwards for a few hundred yards. No outlet could therefore
at any time exist, except along this barrier, and an inspection of its present condition reveals several deep gashes,
through which at one time the surplus water made its
escape. Indeed, some of the oldest inhabitants state that
the name "Lochspouts" was given to it because, in former
times, during heavy floods, its waters spouted across this

ridge at different points.   The truth of this traditional report
is not only consistent with the physical and geological
features of the locality, but supplies a good illustration of
the natural process by which running streams are occasionally known to cut out new channels, and ultimately abandon
their former beds altogether.   Owing to the large amount
of silt washed into this basin, and the gradual lowering of
its outlet by the frictional erosion of the surplus water, the
area of the lake must also have been gradually diminished,
so that it is difficult to estimate its original size.   Immediately prior, however, to human interference with the rocky
barrier, it would not be more than eight acres.   This singular
and, when surrounded by primeval forests, secluded little
lake, was selected by the ancient crannog-builders as a
suitable site for building one of their characteristic island
dwellings, the remains of which have only been recently
discovered.   The starting-point of the investigations now
about to be recorded was the following letter :—

"INLAND REVENUE OFFICE,
CAMPBELTON, 8th October 1879.

"To the Right Honourable
        Sir JAMES FERGUSSON, Bart.

   "SIR,—Would you permit me, a perfect stranger, to bring
under your notice the circumstance that at Lochspoots, on your
estate, there are the remains of an ancient lake-dwelling, which
do not appear to have been ever examined.

   "Lochspoots was formerly of some depth, but within the life-
time of old people the lip of rock which forms its lower rim
was cut with the view of utilising the water of the lake for the
purposes of a walk-mill.   This operation probably reduced the
level about 10 feet, and must have brought the bottom of the
shallower parts to the surface.

   "When on a visit a few years ago to my brother, who is tenant
of this farm, I noticed a mound which I suspected to be the site
of an old lake-dwelling, and on digging into it my suspicion was
confirmed.   My exploration was of the most limited kind ; still

I found a bronze armlet—the metal almost all oxidised—two sling-stones, and two pieces of colouring matter, the one red and the other black. I also ascertained that in cutting a drain a canoe had been dug out of the moss and clay; and on making further inquiry I found it in possession of the previous tenant. I did not measure it, but it appeared small, and to agree with the published accounts of the ruder forms of the canoes found in the Clyde beds.

"As the mound rises above the level of the water, it could be partially examined without much labour or expense; but as the lake water soon finds its way into holes of any depth, no proper or systematic examination could be made without cutting deeper into the ledge of rock that forms the embankment. The rock has already been cut to a depth of 12 or 15 feet, and a few feet more would probably reduce the level below the upper surface of the virgin clay. Fortunately none of the streams that drain into the lake are near the spot, and consequently only a thin covering of lacustrine clay has been deposited over the débris.—I most respectfully remain, sir, your most obedient and humble servant, JAMES MACFADZEAN."

Sir James Fergusson at once forwarded this interesting letter to R. W. Cochran-Patrick, Esq., LL.D., F.S.A., secretary to the Ayrshire and Wigtownshire Archæological Association, with a note requesting him to visit and examine the locality here referred to at his convenience. From letters now before me I find that this preliminary examination of the crannog took place on the 10th of the following November, the result of which was shortly afterwards communicated to me just at the same time that I had received for final revision the proof-sheets of the first article on the Ayrshire crannogs, and so I took the opportunity of recording the discovery by appending a foot-note embracing Mr. Cochran-Patrick's observations. See page 23 of the second volume of the collections of the Ayrshire and Wigtownshire Archæological Association.[1]

[1] As the articles found on the occasion here referred to, as well as the bronze armlet returned by Mr. MacFadzean, have been misplaced, and, in

The time of the year being unsuitable for making an examination of the crannog, owing to the wetness of the locality, it was agreed to postpone further explorations till the following summer.

Meantime the appointment of Sir James Fergusson as Governor of Bombay, and the subsequent return of Mr. Cochran-Patrick as M.P. for North Ayrshire, entirely precluded both these gentlemen from giving their personal attention to the proposed investigations, in which they were so highly interested; and hence the carrying out of them, when a favourable opportunity should occur, was intrusted to me.

Investigations.—It was not till the 28th June 1880 that the weather permitted the work of excavating the mound to be begun, which, however, was then continued regularly during the greater part of the month of July, under the most favourable circumstances. A long course of dry weather made the ground exceptionally suitable for digging; the workmen, with the intelligent forester, Mr. Hopson, at their head, were skilful and thoroughly interested in the investigations; and as to the general management, not only had we the benefit of the able and obliging assistance of Mr. Baxter, factor on the Kilkerran estate, but also the occasional presence and advice of several members of the Council of this Society, among whom were R. W. Cochran-Patrick, Esq., M.P., Sir W. J. M. Cuninghame, Bart., of Corsehill, Colonel Hunter-Weston of Hunterston, J. H. Stoddart, Esq., *Glasgow Herald*, etc. I have specially to mention Dr. Macdonald, rector of the Ayr Academy, who for several days took the

<hr>

the absence of Sir James Fergusson, could not be found so as to be described and figured in the general notice of Lochspouts, I may state that they were as follows, viz., a hammer-stone, a chisel, two bronze armlets, two pointed implements of deer's horn, a granite quern-stone, several bruising-stones, together with a large quantity of bones.

entire supervision of the works and finds. It will be thus seen that the materials of this report are the joint contributions of various hands and various minds, so that the individuality which the writing of it confers upon me must be largely discounted.

Upon my first visit to Lochspouts, I was struck with the smallness of its dimensions; its superficies, according to measurements kindly made by Mr. Brown, clerk to Mr. Baxter, being only 2 acres. Its margin, and indeed its whole area, were thickly covered with long grasses and rushes. On its north side, near the middle portion of the rocky ridge, and a little to the west of the outlet, lay the remains of the crannog, a low circular mound overgrown with coarse grass, and so close to the present margin of the lake that it formed a peninsula easily approached by *terra firma*. I understand, however, that when Mr. Cochran-Patrick visited it in the previous October, the neck of land, now dry, was so soft and boggy that it was with difficulty he got across to the mound.

These observations will be more clearly comprehended by a reference to the accompanying sketch (Frontispiece), taken by a young artist, Mr. J. Lawson, when the explorations were nearly completed. The view is looking northwards. In the foreground are the marshy loch and crannog (the overlying mound being now nearly cleared away), then the rocky ridge extending right and left, behind which is the open valley, with the hill Culdoon, and monument to the late Sir Charles Dalrymple Fergusson in the distance. Along this ridge are seen several hollows, which are supposed to have been formerly outlets; the original or primary one being at the extreme right, while about the middle, and almost in a line with the crannog, is the artificial cutting which forms the present outlet.

Previous to my visit there were no piles detected on the

mound, but after a considerable amount of searching the tops of one or two were observed on its east side, at the bottom of a sluggish channel kept open by the surplus water making its way to the outlet. Guided by these indications and a few trials with the spade, the tops of others were exposed, so that in a short time half the circle was thus traced. After due deliberation, in consultation with Mr. Baxter, who, on behalf of the proprietor, supplied the men and the labouring materials, it was agreed that the only exploration that could then be made, without further cutting of the rock (an undertaking which would involve a large amount of expense), was to clear away the entire mound down to the level of the water. Accordingly, the men were directed to make a broad trench, running east and west, the stuff from which was to be removed in layers, so as to localise, as far as possible, any remains that might be found. When this was finished, another similar trench was made at right angles to the former, after which the four remaining angular portions were removed. In the course of these excavations the following facts regarding the structure and surroundings of the crannog were ascertained :—

1. *Log Pavement.*—About 5 feet deep (measuring from centre of mound), and only a few inches above the level of surrounding water, there was exposed a rude, imperfect, and irregularly-shaped wooden pavement, formed of flattened oak beams. It covered only the central portion of the area contained within the circle of piles, the rest of which was laid with branches and stems of trees. Near the surrounding piles, on the east side, a more carefully constructed arrangement of this wood-work was noticed, consisting of slanting stakes and horizontal beams of various sizes, forming a sort of reticulated and firm flooring, which sloped slightly downwards towards the piles. A similar disposition of the marginal wood-work was noticed elsewhere, especially

on the north-west side, in a line with the gangway to be
afterwards described; but on the lake side of the crannog
the exact mode of its structure was not practically exposed
to view, owing to its shelving below the water, but the pre-
sumption is that it was pretty much the same all round.
On digging beneath this log pavement large beams and
brushwood were generally encountered, but the voluminous
gushing up of water prevented reliable observations from
being made regarding these deeper structures. Occasionally
ashes and charcoal were turned up, and in one spot, near the
centre, and under my own inspection, the men succeeded in
digging downwards more than 2 feet below the log pavement before the water oozed up, in the course of which
nothing was turned out but pure ashes, bits of charcoal, and
large quantities of the shells of limpets and common wilks.
At the bottom of this hole were solid oak beams, apparently
flattened; but no sooner were their surfaces exposed than
the water rushed in and filled the trench. This gave rise to
the conjecture that this under-stratum of remains represented another, and of course an older, period of human occupancy, which also derived some support from the fact that
the surface of the log pavement was on a higher level than
the tops of the encircling piles. It occurred to me, however,
that it was a prepared cavity, and originally intended for
the purpose for which it was evidently used, viz. an ashpit;
and hence, from want of corroborative evidence, the conjecture that the log pavement is a secondary one, and superimposed on the débris of a former dwelling, must for the
present remain *sub judice*. Although portions of mortised
beams were in several instances met with, there were no
remains found of a circle of stockades having transverse
beams, and raised above the log flooring, as was the case at
the Lochlee crannog. Had such a structure existed, it
would have been removed in all likelihood when the lake

was lowered, as the whole wood-work would have been exposed to view. The diameter of the crannog, *i.e.* of the circular area enclosed within the submerged piles, was about 95 feet. No further attempt was made to examine the marginal structure of the island owing to its submerged condition; but the probability is, judging from analogy and the certainty of one circle of piles, that an outer circle exists, with which the former is connected by the usual type of mortised beams.

2. *Hearths.*—Above the log pavement, and a few yards apart from each other, were three circular hearths, each about 5 feet in diameter, formed of flat stones imbedded in a bed of yellow clay, and raised on a sort of pedestal of clay and stones, which varied in thickness from 1 to $1\frac{1}{2}$ foot. One of them, on being demolished, was found to have been built directly over a former stony hearth, with an interval of about a foot. The stuff immediately surrounding them consisted of alternate layers of clay and ashes; and from the number of such layers, indicating collectively a considerable thickness—in one place over 3 feet—it appeared to me that the position of these hearths could not be taken as a criterion of the length of occupancy in the same way as the superimposed series at Lochlee, inasmuch as abundant evidence of the remains of fires were found where no neatly constructed hearth was observed. As will be seen from a glance at the sketch of Lochspouts (see the Frontispiece), they were all situated near the centre of the crannog, but on its southern half, *i.e.* the semicircle farthest from the shore.

3. *Gangway.*—On making a few trial trenches in the space directly between the shore and the crannog in search of a gangway, we could find no indications of wood-work. One day, however, my attention was directed to a portion of the log pavement which looked like a wooden roadway projecting to the margin of the island, and pointing in a north-

western direction, towards a prominence in the trap ridge. Observing, also, that before the lake was lowered this prominence would be the nearest land to the crannog, it immediately struck me that if there was a gangway at all it would be found along this line. Hypothesis was right this time. The adhesive nature of the lake sediment prevented the water from oozing up so quickly as it did on the crannog, so that we were enabled to expose the wood-work several feet below the level of the lake. Close to the crannog the upper beams of the gangway were about 3 feet below the surface of the grass, and fully more below that of the log pavement; but as we neared the shore with the digging they became less buried, and some of the uprights were found even projecting above the ground.

The general plan on which this gangway was constructed appeared to be identical with that adopted by the crannog-builders of Lochlee. Upright piles, singly and in groups, were placed in a zigzag fashion, between which the horizontal beams stretched, fan-like, and so formed a sort of lattice-work, with empty lozenge-shaped spaces between. From one of these holes or meshes, some 5 feet below the surface of the ground, a fine granite quern-stone was extracted. The piles projected some 2 feet or more above the body of the gangway, but there was no appearance of the remains of a platform. The depth of the lower portion of the gangway could not be reached. It would thus appear that at least the transverse beams of the gangway were originally under water—a remark equally applicable to that at Lochlee; and it is highly probable that the primary purpose of this so-called gangway was to supply, on emergencies, a means of secret access to the crannog.

4. *Composition of Mound.*—The surface of the mound was composed of coarse grass, having tough matted roots spreading in a thin layer of soil, which overlay about a foot and a

half of stones and rubbish, in which no relics were found. Below this the materials were of a very variable character; sometimes vegetable mould, stems of grasses jointed like straw, and beds of heather and moss, which could readily be separated into layers; and at other times heaps of ashes and charcoal mixed with quantities of the shells of wilks, limpets, and hazel-nuts. Intermingled with this heterogeneous mass were large and small stones, broken bones, portions of deer horns, and the relics to be afterwards described. Though one or two ashpits, mostly composed of fine ashes, sea-shells, and broken hazel-nuts, were distinctly discernible in the vicinity of the fireplaces, no regular refuse-heap was met with; and the broken bones and horns seemed to be dispersed over the general area of the crannog.

5. *Subsidence of Crannog.*—In discussing the question regarding the Lochlee crannog I had to contend with an element of very great uncertainty, viz., the impossibility of ascertaining how much of the apparent sinking of the crannog was due to the rising of the level of the lake in consequence of the filling up of the bed of the outlet. This doubtful element is, however, entirely eliminated from the problem as it is presented to us at Lochspouts. Whatever alterations may have taken place in the position of the outlet, one thing is certain, that the tendency could never be to raise the level of the lake. Hence, if we can fix on the position of the natural outlet when the artificial cutting was made, the minimum amount of subsidence of the crannog resolves itself into simply measuring the height of this point above the present surface of the log pavement. I use the word *minimum*, because, to determine the actual amount, other two elements have to be considered, both of which tend to magnify the amount of subsidence, viz. (1) How much the surface of the crannog was originally above water; and (2) the amount of lowering of the lake, due to

frictional erosion of the water at the outlet, during the interval between the founding of the crannog and the date of the
artificial cutting of the rock. For the present I entirely
exclude both these elements; so that the solution of the
problem depends on the practicability of ascertaining the
height of the lowest natural outlet above the level of the
log pavement. I believe the primary outlet was at the extreme east end of the barrier, where it disappears into the
hillside. Here is to be seen a large deep opening, naturally
scooped out of the rock ; the lowest portion of which is only
$16\frac{1}{2}$ feet above the present level of the lake. It was, however, found, on measurement, that a lower natural outlet
was just in the site of the present artificial cutting. The
upper portion of the latter is wide, but about 14 feet from
the running water it contracts into a narrow channel with
perpendicular sides, and the sole difficulty is to determine
where nature ended and art began. If we suppose that the
whole of this narrow channel was artificially cut, then the
lake must have been lowered to a corresponding extent.
This, however, may be beyond the mark, as in the course of
time the water itself would make a similar channel. After
repeated and most careful inspections of this spot, I am
inclined to fix the minimum amount of cutting at 10 or 12
feet. Based, therefore, on the lowest estimate, the original
surface of the crannog must have subsided over 10 feet, as
it is now just on a level with the lake water.

RELICS.

No inference worthy of note could be drawn from the
relative position of the relics found on this crannog. They
were interspersed amongst the débris, chiefly around the fireplaces and over the area of the log pavement, at a depth
varying according to their distance from the centre of the
mound, but none more superficial than about 18 inches from

its surface. Though in point of number and variety the general collection is not equal to that from Lochlee, it is scarcely inferior to it in archæological importance. Following the system of arrangement adopted in the latter, I have described the various articles under the several heads suggested by the respective materials of which they are made.

1. OBJECTS MADE OF STONE.

Hammer-Stones.—These implements were in great abundance, forty of which were collected and transferred to Kilkerran House. According to the principle of classification hinted at in the description of those found at Lochlee, which is based exclusively on their shape and the position of the markings, they fall to be arranged in three groups.

First, Two are somewhat flat and circular, about $3\frac{1}{2}$ inches in diameter, and exhibit markings all round the edge.

Second, Three, similarly shaped, have the markings on the flat surfaces alone, and appear to have been held when used with one of the flat surfaces in the palm of the hand.

Third, The rest are more or less elongated, and show wrought surfaces at one or both ends. The largest, made of a fine-grained dolorite, is beautifully polished, tapers slightly towards one end, and measures 7 inches long by 4 broad. A few more were of the same material; and Mr. J. Thomson, F.G.S., Glasgow, informs me that this rock is only found *in situ* at Ailsa Craig, but that water-worn pebbles of it are abundant along the seashore in the neighbourhood of Girvan.

Polishers.—Under this head I classify about a dozen pestle-like implements, notwithstanding that slight pounding markings were observed at the ends of one or two of them, because they are all over so smooth and glossy that they seemed to have been used rather for polishing or

smoothing some soft material, than as hammer-stones. There
are also about a similar number of flat polishers, varying
much both in size and shape, one of which is triangularly
shaped like a modern smoothing-iron.   It measures 5 inches
long, 4½ broad at base, and 1½ inch thick.

Whetstones.—These are also numerous, but it is difficult to
draw a minute distinction between them and the polishers.
They vary in length from 2½ to 6½ inches, and are mostly
composed of hard claystone or indurated sandstone. One
of them, judging from the only fragment which was found,
was manufactured with great care, and had a small hole at
one end for suspension. This fragment, which is here figured
(Fig. 162), measures 3½ inches long, 2 broad, and half an
inch thick.

Fig. 162.- Whetstone (½).   Fig. 163.- -Whetstone (½).

Another is made of fine-grained sandstone, and shaped
precisely similar to the sharpening-stones now used for
scythes.   Its dimensions are 5½ inches long, ¾ inch broad,
and ½ inch deep (Fig. 163).

Funnel-shaped Holes.—Three flat portions of sandstone,
each containing a small hole, opening up on both sides into
funnel-like cavities. The stone here engraved is roughly

circular, about 4 inches in diameter and 1 inch thick.   The
cavity at its mouth is about 1 inch in diameter, $\frac{1}{2}$ an inch
deep, and communicates with a similar one on the other side
by a hole through which a small goose-quill can just pass.
The holes in the other stones are precisely similar in shape,
only the mouth of the funnel in one is one-third larger, and
in the other about as much less; these differences being
entirely dependent on the thickness of the stone (Fig. 164).

Fig. 164.—Perforated Sandstone ($\frac{1}{3}$).

Pebbles.—Of these there were several hundreds found,
scattered all over the island, varying in size from half an
inch to 6 or 7 inches in diameter, the larger of which might
have been used as anvils, others as heating-stones, sling-
stones, etc.

Querns.—Out of eleven quern-stones, almost all of which
were made of granite, only two could be positively stated to
be under ones.   Three of the upper ones were round coarse
lumps, about 1 foot in diameter and 10 inches deep, and of
these two appear to have been unfinished.   One had merely
a cup-shaped cavity on its top, but no hole; and the other,
in addition to the cup, had the central hole partially bored
from both sides.   Neither of them had any marginal hole.

Four were circular, but rather flatter than usual, and measured a little over 1 foot in diameter.

One was oval-shaped and particularly well finished, length 15 inches, breadth 13, and depth 5. The diameter of the funnel at its mouth was 5 inches, and the lower portion of it was lengthened in a line with the main axis of the quern—evidently caused by the friction of the pivot on which it turned round. The smaller end, containing the hole for a handle, was curved downwards, so that its tip was 1¾ inch lower than the under surface of the quern; another striking evidence of the long period the stone had been in actual use.

Fig. 165.—Spindle Whorl (¾).

Fig. 166.—Polished Disc (½).

Spindle Whorl.—One spindle whorl (made of fine sandstone) is 1¾ inch in diameter and ⅝ inch thick (Fig. 165).

Polished Discs.[1]—Two of these interesting objects have turned up on this crannog. One, though wanting a small segment of being a complete circle, is evidently unbroken, as it presents in its whole perimeter a finely cut edge. It is composed of a whitish micaceous stone, quite smooth on

---

[1] Regarding the suggestion that these polished stone discs might have been used as mirrors, see Notes by Mr. Joseph Anderson, *Proceed. Soc. Antiq. Scot.* vol. x. p. 717.

both surfaces, but has no glossy appearance. It measures 4½ inches in diameter, and has a uniform thickness of a quarter of an inch (Fig. 166).

The other, which appears to have been a complete circle, was broken into several portions, two of which have been recovered. These do not fit into each other, but they are so similar in composition, thickness, polish, and size of curvature, that there can be no doubt they belonged to the same disc. The arc of the larger fragment, which is very nearly a semicircle, indicates that the diameter of the completed circle would be 4¾ inches. It is made of a hard, dark, compact stone, highly polished on both sides, and neatly cut at the circumference. It is a quarter of an inch thick at the edge, but becomes gradually a shade thicker towards the centre (Fig. 167).

Fig. 167.—Portion of Polished Disc (½).

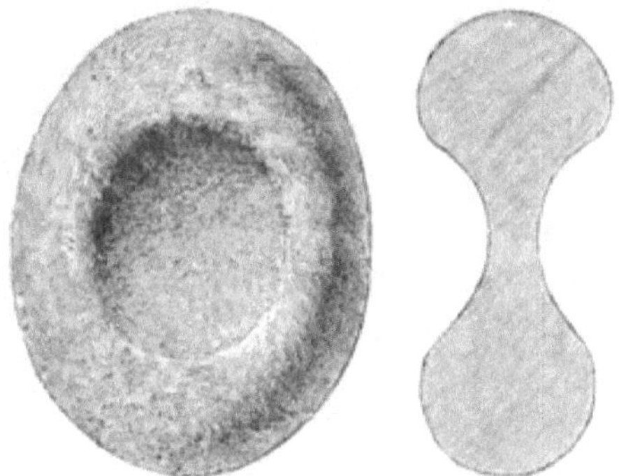

Fig. 168.— Stone Implement (½).

Oval implement with two hollowed surfaces.—This is a smooth oval-shaped stone with a wrought, circular, and cup-shaped depression on each side. Its length is $3\frac{1}{4}$ inches, breadth $2\frac{5}{8}$, and thickness 1 inch. The largest diameter of the depression is $1\frac{5}{8}$ inch, and its greatest depth $\frac{1}{2}$ an inch. It is made of a hard grey trap rock, and, though well wrought all over, is not polished, nor does it exhibit any markings such as are seen on the ordinary hammer-stones, (Fig. 168). See page 56.

Flint Scrapers.—Of these there are two. One, coarsely chipped out of a dark flint, is here figured (Fig. 169). It is roughly circular in shape, and about two inches in diameter. The other is a chip made by a single blow from the outside of a whitened nodule, and is only $\frac{3}{4}$ of an inch in diameter.

Fig. 169.—Flint Scraper ($\frac{1}{1}$).   Fig. 170.—Jet Ring ($\frac{1}{1}$).

Rings of Lignite, etc.—Several bits of lignite or cannel coal were found, some of which showed marks of tools. One small thin bit seems to be the half of a flattened ring, circular on the inside (diameter $\frac{1}{2}$ an inch), but only roughly rounded on the outside.

Ring.—A beautifully polished ring, having a diameter (external measurement) of $1\frac{1}{4}$ inch (Fig. 170).

Armlets.—Portions of two other rings considerably larger, like armlets, one slender, and the other massive and thick.

II. Objects of Bone.

Pin.—A polished pin, length $2\frac{3}{4}$ inches (Fig. 171).

Chisel.—An implement made by cutting a small leg-bone slantingly, so as to present a chisel-like edge.   It is $4\frac{3}{4}$ inches long (Fig. 172).

Awl.—An awl-like instrument. 4 inches long.

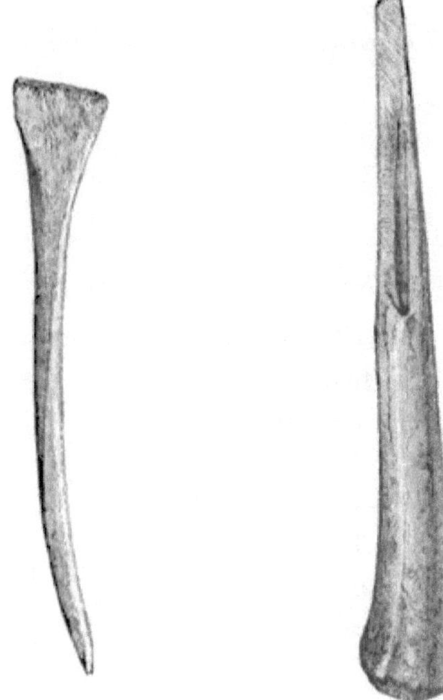

Fig. 171.—Bone Pin (⅓).   Fig. 172. — Bone Chisel (⅔).

Pointed Implements.—Two small-pointed objects, showing marks of a sharp-cutting instrument, and another of a much larger size, being about 6 inches long.

Spatula.—Portion of a flat rib used as a spatula or knife. It is 6 inches long and $\frac{3}{4}$ inch broad.

Knife Handle.—Portion of a shank-bone 2 inches long, hollow in centre, and cut straight across at both ends.

III. OBJECTS OF HORN.

Pick.—Deer-horn pick, made of portion of the horn (as a handle) and the first tine, and much used at point, and also

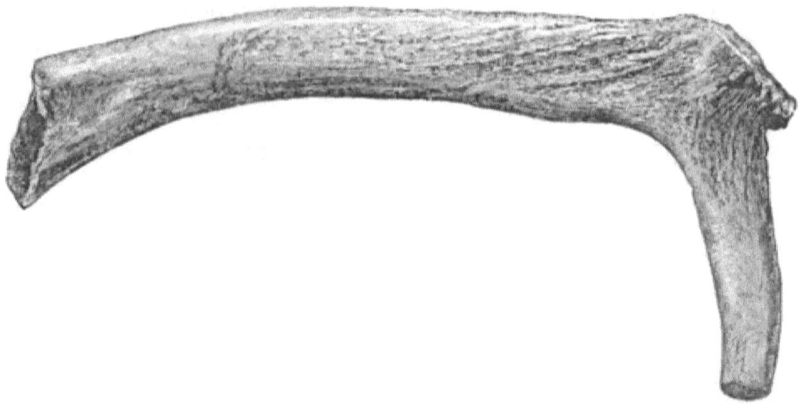

Fig. 173.— Deer-horn Pick (⅓).

on the back, the burr being almost entirely worn off. Length of the handle is 12 inches (Fig. 173).

Club.—Hammer or club-like implement, having the head formed of 3 inches of the root of the horn and the handle of the first tine. This implement is much decayed by long maceration.

Spear-shaped Portion.—This weapon is cut lengthways out of the side of a large red-deer horn, and is 9 inches long and 1½ broad.

Pointed Object.—A slender object, 2 inches long, cut out of a horn lengthways, and sharp at both ends.

Handle.—Cut portion of a tine 3 inches long, and hollowed as if for the handle of a knife.

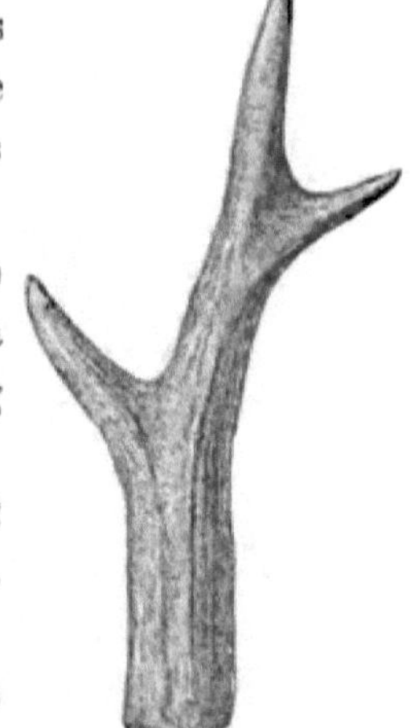

Fig. 174.—Implement of Horn of Roe Deer (⅓).

Pointed Tines.—A few of these show signs of having

been used.   An implement made of the horn of roe-deer is here figured (Fig. 174).

IV. Objects of Wood.

A striking contrast between this collection and that from Lochlee crannog is the paucity of wooden implements.   Indeed, here the only article worth noticing is a slender stave, like that of a milk-cog.   It is 8½ inches long, and the end with the transverse groove is a shade thicker.

V. Objects of Metal.

(*a.*) *Articles made of Iron.*—Articles made of this metal are extremely few.   Besides two portions so corroded that it is impossible to say what they might have been, there remains only one object to be described, viz., a small hand-dagger, much worn and oxidised.   It is 6 inches long, and shows evidence of riveting at the end.

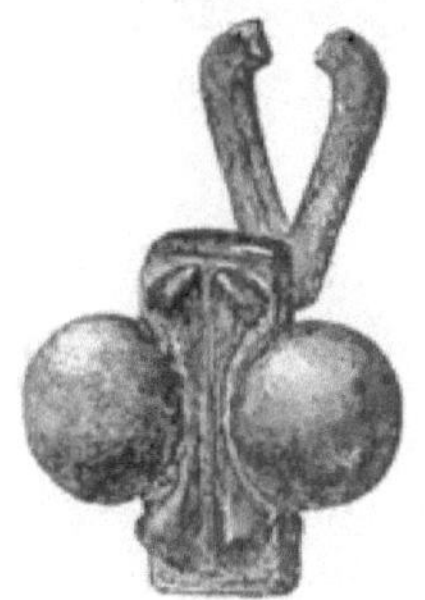

Fig. 175.—Object of Bronze (⅓).

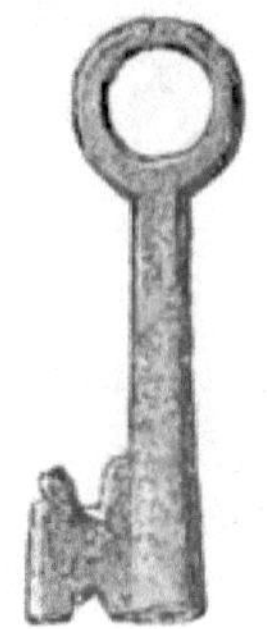

Fig. 176.—Key (⅓).

(*b.*) *Articles made of Bronze or Brass.*—Fig. 175 represents a curiously-shaped ornament, reminding one of the head of a bee.   The parts on its posterior aspect, corresponding to the two circular tuberosities in front, as seen in the drawing, are concave.

Key.—The key which is here figured is 1½ inch long (Fig. 176).

A strong wire, flattened, 4½ inches long, and two small thin plates riveted together, being a fragment of some undetermined object, are all that come under this head, with the exception of the bronze armlet referred to in Mr. MacFadzean's letter, but which has not come into my possession.[1]

VI. Miscellaneous Objects.

Beads.—One small yellowish bead of vitreous paste (Fig. 177). Another ribbed and made of green glazed ware (Fig. 179). Half of another, very similar to the last both in colour and composition, but considerably larger, and having the hole contracted about its middle by a raised circular ridge (Fig. 178).

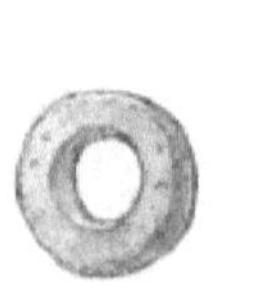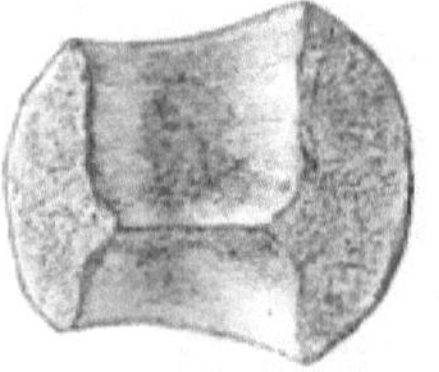

Fig. 177 (¹⁄₁).   Fig. 178 (¹⁄₁).   Fig. 179 (¹⁄₁).

Beads.

Pottery is more abundantly represented than at Lochlee, though of a similar character, and in both crannogs portions of Samian ware have been found.

Fig. 180 represents portion of a bowl of Samian ware, showing its characteristic moulding, the festoon and tassel, commonly called the egg-and-tongue border, and portions of

---

[1] In case K, in the York Museum, which was constructed in 1872-3, to hold the specimens of Roman Metal Work, Implements, and Ornaments of Bone, Jet, etc., are:—Four keys almost identical with that figured above; several articles of bronze of a similar type with the object represented by Fig. 175; a small circular bronze brooch with transverse grooves like Fig. 241; three small bifurcated objects like Fig. 240; harp-shaped fibulæ, like those from Lochlee; besides many bone pins and combs, jet rings, beads, etc., all of which are wonderfully like the corresponding articles found on the crannogs.

the ornamental figures with which it was adorned.   Its fine
texture is of a uniform reddish colour, but the glaze has a

Fig. 180.—Portion of Samian Ware (½).

redder tint.   The diameter of the mouth of this vessel would
be between 6 and 7 inches.

Fig. 181.—Pottery (⅔).

Three other fragments of similar ware, but of a more slender build, were collected. These might all belong to the

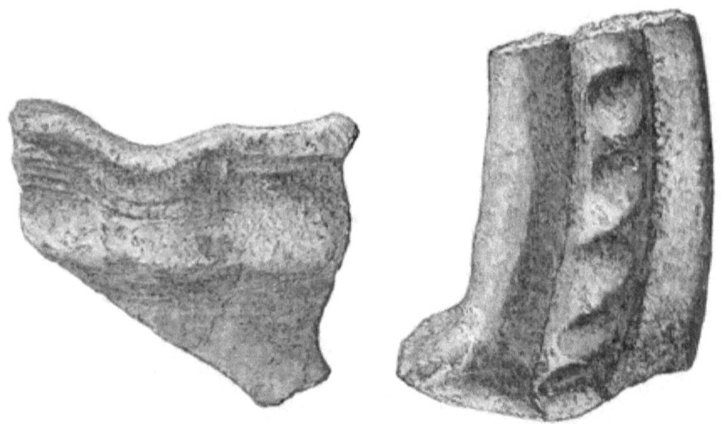

Fig. 182.—Pottery (⅔).　　　Fig. 183.—Handle of Vessel (⅔).

same vessel, and they presented no appearance of ornaments.

Figs. 181 to 185 are illustrations of another kind of

Fig. 184.—Pottery (⅔).

pottery. It is of a light colour, feels soft to the touch, and is mixed with coarse sand. Its thickness is somewhat variable, but rarely exceeds ¼ of an inch. The fragments represented by Figs. 182 to 184 show small patches of a yellowish-green glaze.

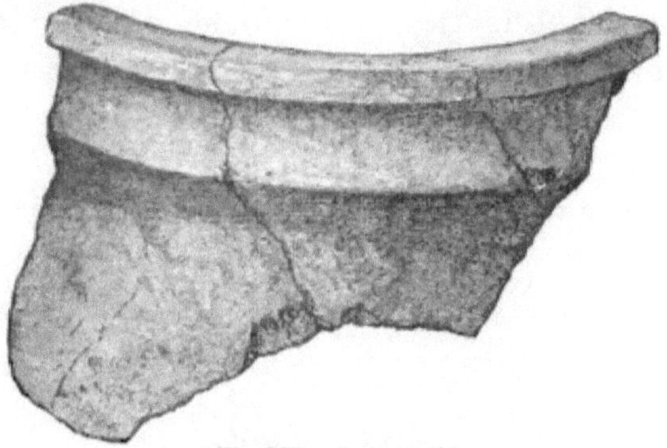

Fig. 185.—Pottery (⅔).

Fig. 186 represents another class of pottery very different from the latter. It is nearly ½ an inch in thickness, and is altogether more massive, but contains no coarse sand, and its colour externally is a dull black.

Fig. 186.—Pottery (⅔).   Section of Fig. 186.

Organic Remains.—At his own request, a selection of the bones and horns collected during the investigations was forwarded to the late distinguished and much lamented Professor Rolleston of Oxford, for examination and comparison with those from Lochlee, but unfortunately, owing to the state of his health, he was unable to make a report. I may state, however, that the *osseous remains* were very

similar to those from Lochlee. The bones of the sheep,
amongst which was an entire skull, were proportionately
in greater numbers than either those of the pig or ox.
Horns were very abundant, but included only those of the
red-deer and roe-deer. Judging from the amount of the
remains of shell-fish (*Lit. littorea, Patella vulgata,* and
Trochus), they must have been largely consumed as food.[1]

SECTION III.

Notice of a Crannog at Barhapple Loch, Glenluce, Wigtownshire.

(By the Rev. GEORGE WILSON, Glenluce, C.M.S.A. Scot.)

Barhapple Loch, on the farm of Derskelpin, lay a little
to the south of the road from Portpatrick to Dumfries, just
beyond the fourth milestone east from Glenluce, between
two round hills called Derlauchlin and Barhapple, and about
285 feet above the level of the sea. The water-parting is at
Barhapple Hill. The loch was about 1500 feet long, and
1000 feet broad, surrounded by deep peat bog, except on
part of the east shore where it touched Barhapple, and
rested on a bottom of deep soft peat. Although the water
was only a few feet deep, its black colour and the inaccessible nature of the shore on the west side prevented the discovery of any trace of lake-dwellings. It was drained in

---

[1] Since writing the above I understand that the natural basin of
Lochspouts is about to be converted into a reservoir for supplying the
town of Maybole with water, and that, in order to make it suitable for
this purpose, according to the engineer's report, it will be necessary to
clear away the whole of the lake sediment, including the crannog, at an
expense of some £900. As no explorations directed from an archæological
point of view could be more satisfactory than these contemplated operations, we may expect, in the course of their execution, to find not only
additional relics that may have dropped into the surrounding lake, but
to secure absolute accuracy regarding several doubtful points, such as the
dimensions and mode of structure of the island, etc. See Appendix.

the autumn of 1878, and in November of that year Mr Shearer, the tenant, told me that a small round patch of logs and stones had become visible.

On the 15th of October 1880, our President, the Earl of Stair, assembled a party to explore the crannog. There were present with him Admiral Sir John C. Dalrymple-Hay, Bart., M.P., and Sir Herbert E. Maxwell, Bart., M.P., two of our Vice-Presidents; the Hon. Hugh Dalrymple, Mr. J. Pendarves Vivian, M.P., Mr. Vans Agnew of Barnbarroch, M J. Leveson Stewart of Glen Ogil, with Mr. R. W. Cochran-Patrick, M.P., and myself, the Secretaries of the Association. Our digging was stopped at a depth of two and a half or three feet by the influx of water, yet we found a good deal to interest us. This lake-dwelling, so far as explored, consists mainly of piles and platforms of wood, with rough stones at some points. It is about 280 feet from the west shore, but the gangway had run about 550 feet to the east shore at the foot of Barhapple, where there is hard ground. It is surrounded by a row of oak piles, enclosing a space 175 feet long from north to south, and 127 feet broad, and rounded at the angles. While the digging was going on Sir Herbert Maxwell took these measurements for me, and Mr. Vivian walked round on the soft peat and counted the piles in the outer row, of which 134 were visible. There is a slight gap at the west side, and a larger one on the south side, with the piles on each side of it more thinly set. An irregular line on the Plan marks off a part of the enclosure on the east side, which is about 9 inches higher than the rest, and is the only part that can be walked upon with ease in ordinary weather. After heavy rain the whole is still inaccessible, owing to the imperfect outfall of the drainage.

Thirty-one feet from the outside piles towards the south-east, there was a layer of rough, large stones, marked B on

Plate III., about 15 feet long from north to south, and 11 feet broad. Seventeen feet farther north, and 18 feet from the east side, there was a spade-shaped platform, with the convex end to the north, about 26 feet in length and breadth. The Plan shows its appearance in February 1879, with several pieces of wood flooring towards the east side, and a layer of large rough stones at A. In October 1880 some of the logs had rotted away, and others were pierced through by the shoots of the marsh plants, which are gradually covering the partially drained area. Thirty feet to the west of A, there was a circular layer of rough stones about 10 feet in diameter, surrounded by several rings of piles. On removing some loose dry peat on the east part of A, we found a floor of oak logs, laid north and south, 10 feet 6 inches in length and 8 feet in breadth. The surface was somewhat flat; but this may have been caused by exposure to the weather. The interstices were closely packed with white clay and the sphagnum moss, so common in our bogs, with a few stakes driven between them. At the west or inner side of this floor there was a log 13 feet 6 inches long, 1 foot broad, and 8 inches deep. Beyond it was a layer of large rough stones from 9 to 12 inches deep, which had been disturbed by some idle visitors, so that its exact extent cannot be given. Under the stones was a thin layer of peat, then a log floor resting on clay and stones, and under that a second floor, the parts of which were sloping. Under the large oak log already mentioned lay a few birch logs sloping towards the north-west, and covering at the left side one angle of a frame 6 feet 6 inches square, made of four oak beams, that on the south-east side having two square-cut mortise holes, measuring 6 by 5 inches, and 4 feet apart, and that on the opposite side having one mortise hole with a piece of the upright still in it. In the angle between this frame and the south end of the large log there was a circular hearth of rough stones bedded in clay, and a similar hearth beyond the north-west angle, where there seems to have been another square frame without mortises. There were several inches in depth of ashes, with charred wood, and fragments of bone too small and wasted to indicate what animal they belonged to. West of the second hearth the following section was noted in descending series :—

  (*a.*) Rough stones, 9 inches.
  (*b.*) Peat, 12 inches.
  (*c.*) Ashes, 5 inches.
  (*d.*) White clay, 3 or 4 inches.
  (*e.*) Ashes.

Under the floor first described there was a layer of smaller sticks and branches of oak, hazel, and birch, and at the north-east we found under the branches a layer of the common bracken, *Pteris aquilina*. The influx of water prevented further examination, but at different places the spade struck on logs which could not be seen. The wet state of the peat, ashes, and clay made exact search difficult. Near the second hearth we found a long rude whetstone, a hammer-stone of water-rolled quartzite pebble, a fragment of smoothly-worked wood, 3 inches long, 2 broad, and $\frac{1}{2}$ an inch thick, which may have been part of a ladle or large spoon, and a small branch like one's little finger, rudely pointed, and with an untrimmed bent head. When unpacked at the Museum, these pieces of wood had gone to pieces.

A trench cut from the hearths to B showed logs and stones under the stone floor there, but not in any regular order. Under the stones, at C, we got two broad pieces of oak about $4\frac{1}{2}$ feet long, which may have been parts of a canoe.

Near the beginning of the gangway, at the end of a log, there rolled from a labourer's spade a ring of unevenly polished cannel coal, which is shown in Fig. 187, full size.

The piles are pointed, and show the axe-marks distinctly.

Two or three branches, 2 inches thick, had been severed by a single cut. The piles are from 6 to 8 inches thick, but I saw one a foot thick. One which was pulled up was 5 feet long. The Plan shows the radiating and curved arrangement of the piles.

Fig. 187.—Ring of Cannel Coal (⅓).

At the south-east of the crannog, a few feet from the edge, two piles 6 feet apart show where the gangway entered. Two or three are seen farther off, then about twenty at a place where the gangway seems to have widened to nearly 12 feet, and beyond these are two other pairs, the last being about 100 feet from the shore. Beyond that the piles have rotted away through exposure to the weather in dry seasons. There are decayed remains of timber at various places round the shore.

While we were digging at the crannog, Sir Herbert Maxwell, who is an experienced observer of lake-dwellings, explored the whole circuit of the loch, and reported that he had found some logs laid like a corduroy road. I did not see them at the time, and when I went back frost and flood had hidden the traces of them. At the letter c I have indicated pretty nearly the spot where they were seen. Perhaps another platform was there.

In April 1881, when verifying some details, I observed a few piles at the point marked D, between the crannog and the north shore, and reached them with difficulty. The nearest is about 120 feet from the shore, and is the first in a straight line of four piles, set at distances of 6, 10, and 8 feet, with two others 6 and 7 feet to the left, nearly opposite the second and third. At E I have marked the probable position in the peat bog of an object described by me in "Notes on the Crannogs and Lake-Dwellings in Wigtown-shire," in the *Proceedings of the Society of Antiquaries of Scotland*, vol. ix. page 337,—"*Barhapple Loch*, four miles east of Glenluce, close to the coach road.—James M'Culloch, one of my deacons, told me that about the year 1842, in cutting peat about 40 yards from the west side of this loch, he came on a circle of stakes (about a dozen) from the thickness of the arm to that of the leg, and about 5 feet long; the heads at least 2 feet below the surface. The stakes were of hazel, pointed by four axe-cuts, 3½ to 4 inches broad, and some of them 5 inches long. The circle was cut away at two times, and was at least 5 feet in diameter; coarse branches were twisted among the stakes like wicker-work. No trace of clay." In 1871 I reported this as indicating that some dwellings might yet be found in this loch. It seems to have been a *marsh*-dwelling, like some of those found near lakes in Switzerland.

The crannogs were probably used as places of refuge, although they may also have been occupied constantly. There is often a fort on the top of some neighbouring hill, to which the lake-dwellers may have gone when the lochs were frozen and the crannogs open to invasion. We have an example of this at Machermore, Glenluce. The two round hills between which Barhapple Loch lay have both been

ploughed, and show no trace of fortification or dwellings.
But beyond Barhapple, and half a mile eastward, on the
farm of Barlea, a small knoll south of Barfad rises out of
the bog like a peninsula. It is nameless on the Ordnance
Survey Maps, but on an old map of Blairderry and Barlea,
which must be above a hundred years old, it is called *Drum-
earnachan*.   There are traces here of an old village or
settlement, although it has been partially ploughed.   At
the lowest part of Barfad there is a ring of turf and stone
17 by 16 feet in diameter. 138 feet to the south are the
remains of a wall or breast-work 126 feet long and 12 broad.
Beyond it several foundations are seen in a straight line
north and south.   At 96 feet is the bottom of a cairn 30 feet
long and 22 broad, and 40 feet to the left of it a roughly-
paved circular floor, 6 feet in diameter, which has been saved
from the plough by having a large boulder rolled on to it.
Thirty-six feet beyond the cairn is a 9-feet circular foundation of stones; 26 feet farther on an oval lying across
the line, 15 by 13 feet; 8 feet farther on, an 11-foot ring;
59 feet beyond that, a small circular patch of stones; and
another, 45 feet farther on, with a low grassy cairn 10 feet
in diameter, 36 feet off at the west. Sixty-two feet south-
east from the last foundation in the straight row is a circular
turf and stone ring, 10½ feet thick, 3½ high, and 48 feet in
diameter over all, with the entrance-gap at the *south-west*.
On the 6-inch Ordnance Map it is marked "site of cairn,"
but I have never found any one who had heard of a cairn
there.   Part of the enclosed space is somewhat stony, and
the position of the entrance-gap is peculiar, all the others I
have seen or heard of having it at the south-east. Many
years ago, the late tenant, Mr. M'Ilwraith of Kilfillan, asked
me to go and see this ring, because he thought it had been
surrounded by two oval rows of earth-fast stones. I went
and made careful measurements, with this result, that the

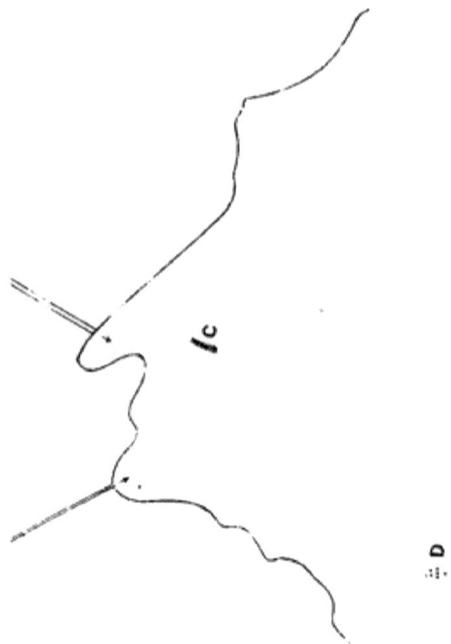
c
D

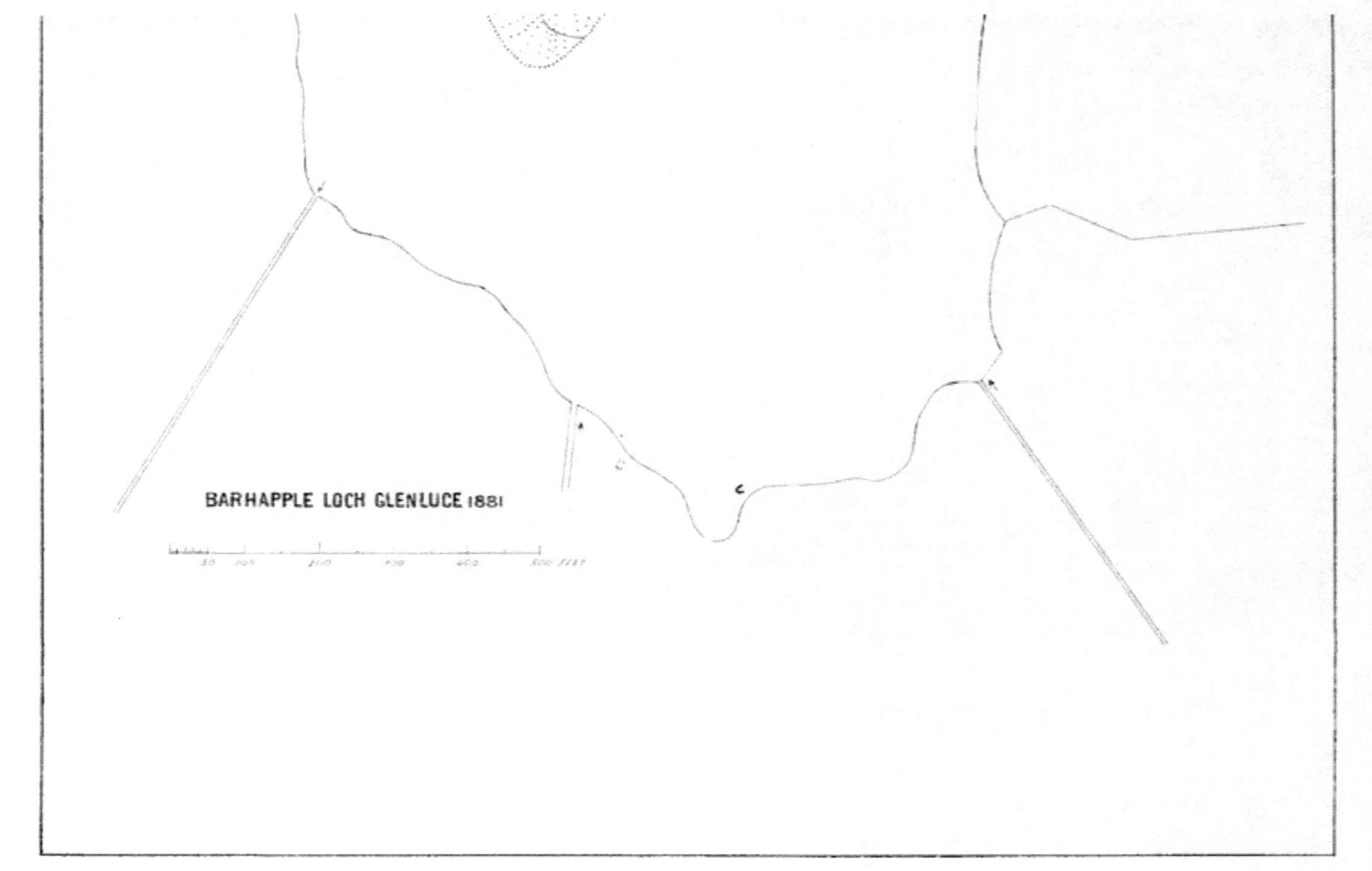

BARHAPPLE LOCH GLENLUCE 1881

stones *may* have been arranged in order, but there has been too much disturbance by the plough to make this more than a guess. For a long time I regarded such rings as small forts; but have lately begun to think they may have been places of interment. I have heard of three instances in which the plough, in levelling down such rings, turned up crocks of coarse pottery, not in the enclosed space, but in the rings themselves. The attention of observers elsewhere is called to this fact.

Half a mile due north from the Barhapple crannog, passing Knockiecore, Barrel Hill, and Derniemore Hill on the left, and Tamricroach Moss, Derhagie Hill, and Blairderry Hill on the right, just beyond the old military road, we reach a low rocky hill surrounded by a peat bog, which unfortunately has lost its ancient name, and is called from its broad shape the Braid Hill. It is on the farm of High Dergoals; and Mr. Dougan, the tenant, told me that many years ago he found, in cutting peat at the south side of it, at a depth of 4 feet, three or four stakes, apparently of oak, 3 or 4 inches in circumference, and pointed by a single cut. The higher ground is rocky and uneven, and scattered over it are the remains of several small cairns and rings. At the west end is a 10-foot ring, a cairn with the remains of a stone grave in the centre, and beyond it two others lying east and west, with a foundation between them, 27 by 14 feet, with the corners much rounded. Towards the middle there are two circular foundations, three others on the north slope, three on the south, and three more at the east end, all so indistinct that it is difficult to say whether they have been huts or cairns. On the slope at the east end there are two rings. It is impossible to know whether either of these sites has been occupied by the Barhapple lake-dwellers. There are no others near it, although there are several other ancient village sites in Glenluce, some of which I hope to describe in a future volume. There have been four other lochs in Old Luce parish with crannogs. The frequent occurrence of the syllables *der*, *dir*, or *dur*, in the names of the places near Barhapple, shows that long ago they were clothed with trees. Here is a topographic rhyme, by some unknown native bard, communicated to me by Mr. Thomas M'Cormick, farmer at Mindork, in Kirkcowan :—

> " Knocketie and Knockietore,
> Laniegoose and Laniegore,
> Dirnefuel and Dirniefranie, wee
> Barsolas and Derruagie."

Section IV.

Notice of the Excavation of a Crannog at Buston, near Kilmaurs.

Discovery of the Crannog.—About half-way between Stewarton and Kilmaurs there is, on the farm of Mid Buston, the property of the Earl of Eglinton, a shallow basin, now converted into a richly cultivated meadow, but which formerly, as represented in Bleau's *Atlas*, formed the bed of a lake of considerable size called Loch Buston. Within the recollection of the present generation this area was a mossy bog in summer and a sheet of water in winter; and about fifty years ago, when the present tenant, Mr. Robert Hay, came to reside on the farm, there was a small mound or island situated about its centre, locally known as the *Swan Knowe*, on account of the numbers of wild swans that formerly used to frequent it. When subsequently engaged in reclaiming the bog, Mr. Hay states that as many as thirteen cart-loads of timber were removed from the "Knowe," and he distinctly remembers that, in consequence of the difficulty of detaching some of the beams mortised into others, his father then made the remark, "there maun hae

been dwellers here at ac time." He also states that until
the land was thoroughly redrained, some five years ago,
there was still a considerable mound to be seen; but at the
beginning of December 1880, when I first visited the locality,
there was hardly any elevation to distinguish it from the
surrounding field. Notwithstanding Mr. Hay's knowledge
of the structure of the " Knowe," which he supposed to have
been erected by one of the old Earls for the purpose of facili-
tating the shooting of wild-ducks—a purpose for which it
had frequently served himself,—the merit of detecting here
the ruins of an ancient lake-dwelling is due to Mr. D.
M'Naught, schoolmaster of Kilmaurs. The history of the
discovery is most interesting, and reflects much credit on
the discoverer; but the story is best told by himself.
Having a faint recollection that Mr. M'Naught was one of
a group of critical sceptics who visited Lochlee while the
investigations there were in progress, and maintained that
the crannog was merely the site of an old " whisky still," I
was curious to know the circumstances and exact process of
ratiocination which had now actually culminated in placing
him in the position of being a discoverer in this same line
of research; so, after the importance of the crannog had
been established by some valuable " finds," I wrote a note
asking if he would kindly oblige me by a written statement
of whatever information he could supply on the subject.
The following is his reply :—

" KILMAURS, January 15th, 1881.

" DEAR SIR,—I have much pleasure in replying to yours
received this morning.

" About five years ago, when engaged in levelling the large
drain that passes Buiston crannog, I passed over the very spot,
but being utterly ignorant on the subject I noticed nothing
peculiar. When passing through the stackyard on my way home
I noticed the old beams, but on being told that they were from
some old house I thought no more of the matter. The subject

had so completely escaped my memory that even when I had seen the Lochlee beams they failed to recall what had formerly arrested my attention at Buiston. My scepticism at Lochlee arose from the fact that I failed to trace the shape and construction of the crannog as detailed in *Chambers's Encyclopædia*, which was the only authority then at my disposal.

"I never heard anything more of the Buiston crannog till the week of the discovery. Talking with one of the farmers in my own house, the conversation turned on furniture, when bog-oak was mentioned. He remarked that there was as much lying in Buiston stackyard as would stock the parish. At once I remembered what I had formerly seen, and though the recollection was hazy, on afterthought I felt almost sure that I had noticed mortised holes, and that *the beams were identical with those I had seen at Lochlee.* Next day, as soon as I had closed the school I went up to the farm. Mr. Hay was inclined to pooh-pooh the matter, and said that the place was 'juist a timmer house ane o' the auld Earls had put up to shoot deuks.' Going out to the stackyard I found that the ricks had been built on the old timber, which made excellent 'bottoms.' I looked about for an odd bit, and did eventually get a splinter, but not sufficient for identification. After getting rid of the old man, his youngest son and I set to work at the bottom of one of the ricks, and pulled one of the beams so far out as enabled me to saw off the mortised joint. This I sent to the *Standard* office, where you saw it on the Saturday morning following. I then went down to the site of the crannog, but it had become so dark that I had to feel my way. I eventually kicked against something which seemed to be an upright sticking through the soil. I went up next morning early, and when I had seen the three uprights afterwards pointed out to you, and the mortised beams stuck in the side of the drain, I no longer had any doubts. I therefore at once wrote to Mr. Cochran-Patrick, and penned a cautious intimation for the *Standard*, which the editor accepted on trust from me. You know the rest.—Yours truly,

"D. M'NAUGHT.

"DR. MUNRO."

On the afternoon of the Saturday referred to in the above letter (December 4th, 1880), I accompanied Mr. M'Naught to the *quondam* "Knowe," and in a short time, by a few tentative diggings, the existence here of the remains of a crannog was put beyond a doubt. Our Secretary, R. W. Cochran-Patrick, Esq., M.P., who had already been communicated with, then brought the matter under the notice of the Honourable G. R. Vernon, Auchans, as Commissioner for the Earl of Eglinton ; and after due deliberation it was agreed to make an immediate investigation of the crannog on behalf of the proprietor. Accordingly, on the 10th December 1880, six men were started to work in presence of Mr. Vernon, Mr. Cochran-Patrick, and several ladies and gentlemen interested in the discovery. It is needless to describe the subsequent management of the excavations. The peculiar and absorbing interest excited by the variety of the finds during the first few days soon developed the true spirit of inquiry among all concerned, and even the old and highly-respected farmer gave up his long-cherished theory of the "duck-shooting," and ultimately rendered valuable aid by protecting the trenches from the prying curiosity of the general public, and picking up relics from the stuff wheeled out, which became visible by long exposure to weather and heavy rains. By general consent, at least *nem. con.*, I was appointed custodier of the relics ; and now, acting on the old saying that possession is nine points of the law, I have assumed the rôle of historian.

Method of Excavating.—The excavations were commenced by making an explorative trench through what appeared to be the centre of the crannog, following as a guide the long diameter of the lake basin. This trench was from 2 to 3 feet deep, and about 5 feet wide, and its general direction lay in a line running from N.W. to S.E. The débris was wheeled sufficiently far not to cover the probable area of the island, and carefully examined, but nothing of importance was found, except a small spindle whorl (Fig. 196), and a fragment of a quern-stone, till the trench reached the

southern margin of the crannog. Here, after the tops of
a few upright piles were exposed, a large beam was encountered, lying right across the trench, beyond which the
stuff turned up from the bottom consisted almost entirely
of broken bones and ashes. This was at once recognised
as the wished-for midden, and its discovery at this early
stage was fortunate, inasmuch as its examination would
soon decide, with a trifling outlay, the quality of the crannog
as a relic depôt. To this, therefore, attention was exclusively devoted, till the severity of the weather compelled us
to abandon working altogether. The depth of clay and soil
above the midden was about $2\frac{1}{2}$ feet, and after removing
this, its remaining contents were wheeled to a separate
place, so as to facilitate a more careful inspection after
exposure to winter weather. The large number of rare
and valuable relics discovered during the ten days the men
were thus employed induced the Earl of Eglinton to sanction
a further outlay in the prosecution of these researches; and
it was then agreed that nothing less than the removal of the
débris over the whole area of the crannog would satisfy
archæological demands. The tenant also very kindly con-
sented to leave this portion of his field untilled, so that
there was no necessity to resume work till the weather
became really suitable for such an undertaking.

Early in April very dry weather, though cold, set in,
and on the farmer representing that more favourable circumstances for digging could not be expected, the investigation of the crannog was resumed.

While clearing out the refuse-heap, the position of the
surrounding piles immediately to the left of the original
trench was readily ascertained to be arranged in three or
four circles. With these as guides, it was an easy matter
for the workmen to clear away the soil right round the
central portion of the crannog without the necessity of

constant supervision. The surface soil, which consisted of fine clay, varying in depth from about 6 inches at the centre of the mound to 2 feet beyond the outer circle of stockades, was first wheeled away, and, as no relics were expected here, there was no time wasted in searching for them. Afterwards the dark heterogeneous under stratum of débris was carefully removed from above the wood-work and examined, though not with the same care as the contents of the refuse-heap. Here, however, a few important relics were discovered, among which are an ornamented gold spiral finger-ring, a small earthen crucible, and some fragments of pottery. Having completed this broad annular trench, the débris remaining on the central portion was taken away, but, contrary to expectation, nothing was found in it beyond the evidence of a few fireplaces, some slag, and one or two large wooden pins.

Structure of Island.—Notwithstanding the havoc committed on the wood-work of the crannog by long exposure to atmospheric agencies before it finally sunk under the protective influence of the muddy water, and subsequently, by the ruthless hands of the agriculturist, there still remained sufficient materials to give one not only a general, but particular and instructive notion of the mechanical principles on which the island was constructed. Its substance, as far as could be ascertained by digging holes here and there, was made up of layers of the stems of trees, chiefly birch; intermingled with which were occasionally found various other materials, such as brushwood, heather, moss, soil, and large stones. Penetrating deeply this heterogeneous mass, towards its margin, were numerous piles, forming a series of concentric and nearly circular stockades, which were separated from each other by an interval of 4 or 5 feet. On the south side there were four distinct circles to be seen, but on the north only three could be detected, as the

Fig. 188.—General View of Buston Crannog, looking northwards. The water in foreground marks the position of Refuse-bed. (*From a Photograph by Mr. Laurie.*)

third outermost appeared to have merged into the external
one; and, in accordance with this diminished number of
circles, the breadth of the stockaded zone also diminished.
The piles in the inner circle, which were strongly made, and
showed evidence of having been shaped and squared by
sharp-cutting instruments, were uniformly arranged at a
distance of from 4 to 5 feet, and enclosed an area more of
the form of an ellipse than a circle (measuring 61 feet
by 56), while those in the second and third circles were
more irregularly, but generally more closely, set. All these
uprights (except a few on the north side of the inner circle)
were linked together by horizontal beams having square-cut
holes, through which the upper ends of the piles passed.
The horizontal beams were arranged in two ways. Some
lay along the circumference and bound together all the
uprights in the same circle to each other, while others took
the radial position and connected each circle together. Some
of the latter were long enough to embrace three circles, and
when this was the case I have noticed that the upright in
the middle circle was sometimes firmly caught in a deep
cut in the transverse, instead of passing through a mortised
hole (see Fig. 190). Although the uprights in the inner
circle were not linked together circumferentially along the
whole course of the horizontal beams, the particular construction of the log pavement on the north side rendering
this unnecessary, every one of them had a radial beam,
directed from within outwards, which kept it from yielding
to lateral pressure. This purpose was equally well served
in several ways; sometimes the inner end of the radial beams
pressed tightly against the upright, at other times the former
projected half-way into the log pavement, where its end
was firmly fixed by a thick pin passing through it into the
under structures of the island, and its middle contained
either a notch or mortised hole for holding the latter in

position. The external ends of these radial beams were occasionally observed to be continuous with additional strengthening materials, such as wooden props and large stones.

The main object of the whole of this elaborate structural system was to give stability to the island, afford fixed points on its surface, and prevent the superincumbent pressure of whatever buildings may have been erected over it from causing the general mass to bulge outwards—objects which appeared to have been most effectually attained.

The piles in the outer circle were merely round posts, smaller and more closely placed than those in the inner circles, being sometimes only a few inches apart, and appeared to have been bound together by a transverse rail, into which their tops were inserted after the manner of a hurdle. Portions of these rails, pierced with holes, were found at the south-east side, though none actually in position; so that the inference that the outer stockade was intended as a fence or bulwark seems quite legitimate. In support of this view I may state that nowhere along its course were the piles connected together by horizontal beams, either circumferentially or radially, nor did they penetrate deeply, so that for giving stability to the island the outer circle would be of little use.

Log Pavement.—Like the other crannogs examined by me, this one also had its central portion roughly paved with wooden beams like railway sleepers. On looking at these beams carefully it was observed that many of them, especially those made of oak, had also holes at their extremities, and that the plan of being linked and fixed together by stout wooden pins was by no means peculiar to the marginal portion of the crannog. Here, however, they lay mostly in a radial position, and on the south side; some were distinctly seen to be joined with the uprights in the inner circle with one end, while the outer, which pointed to the centre, was firmly pinned to the wood below. In several parts this general network of large beams was covered over by a pavement made of small round logs, mostly of birch, and placed close together, but, being soft and easily removed, I could not be certain whether or not it extended over the whole area. If so, it must have been a secondary pavement formed after the crannog was inhabited, as marks of fire, with slag and ashes, were found in two or three places lying immediately on the large oak beams below it.

On the north side of the crannog the uprights in the inner circle were not linked together circumferentially by horizontal beams, because (as I have already remarked) the different structure of the log pavement here rendered this plan unnecessary. The reason of this was, that on this side a considerable segment of the log pavement was built up, for a depth of 2 feet or so, of several layers of those round logs of soft wood, laid transversely to each other, and carefully arranged flush with the outer edge of the uprights, so that the only direction in which the latter were free was counteracted by the radial transverses alone (see Fig. 190).

The space between this portion of the log pavement and the next circle of stockades was filled up with layers of turf and moss, the depth of which corresponded with that of the built-up edge of the log pavement. After removing the turf and moss from this space in one or two places, we came on the wood of the island, which here consisted entirely of birch-trees with the bark on, and looking as fresh as if they had been recently cut. The heather and moss also looked quite fresh, but soon, after exposure to the air, everything turned black. (See Plate IV.)

Remains of Dwelling-house.—Over the area of the log pavement there were here and there the remains of large uprights, which appeared to have been used as supports for some sort of dwelling-house. On the north side, a few feet from the margin of the log pavement, there were three or four of these, as if forming another circle, one of which I extracted with difficulty and found it to be 8 feet long, 7 of which were imbedded in the structure of the island. It was neatly formed of a rectangular shape (10 inches by 6), and its downward end was cut and pointed as if for insertion into a mortised hole. The centre of the log pavement was occupied by a mass of ashes, charcoal, and stones, forming a bed about 2½ feet thick, being nearly the entire depth of the mound above the wood-work, and a little to the west of this, and situated between two large square-shaped uprights, there was a thin bed of charcoal and burnt straw, together with some flat stones covered with a quantity of slag. On the east side, near the circle of piles, conclusive evidence of another fireplace was observed, but no well-formed hearths were anywhere met with.

On tracing the inner circle of stockades all round, it became evident that they formed part of some sort of enclosure. On the south-east side were two well-shaped rectangular uprights, about 2 feet 6 inches high, and 4 feet apart, firmly mortised into a well-constructed wooden flooring. These, as will appear from the sequel, formed portions of the door-posts of the entrance to the area of log pavement. Continuous with them, on the east side, and in the line of the inner circle, some of the intervals between the uprights were actually found to contain the remains of a composite wall of stone and wood. The space between the second and third piles, counting from the doorway, was thus filled up. At the base there were two layers of rectangular stones, then a flat beam of oak laid horizontally, then three layers of thin flagstones, well selected for size and shape, then another oak beam similar to the first, and finally other three layers of flat stones. This wall had partially fallen over, but the

relative position of the respective layers was still retained,
and showed that when standing it would be about 3 feet
high (see Fig. 189).   The adjoining space, next the doorway,
had two layers of stones at the base, and then a beam, but

Fig. 189.—Eastern portion of Crannog, showing surrounding Stockades and portion of Log Pavement. The Sign-board marks the position of Canoe. (*From
Photograph by Mr. Lawrie.*)

the rest was wanting. There were no further remains of
a decided wall met with, though stones were abundantly
encountered all over the area of the crannog. As all the
uprights in the inner circle appeared to have been worn or

broken, there is no evidence to show what their former
height was, but as they now stand, they are not only different in shape, but considerably taller than those in the
second and third circles, which are all shorter and more or
less pointed.

Directly facing the door-place, but 13 feet farther out,
and nearly in a line with the outer circle of stockades, there
was a large rectangularly-shaped beam 11 feet long, containing two mortised holes, one at each end, and having an
interval of 8 feet 6 inches between them. This beam lay
close to two massive uprights which projected about 2 feet
above the surface of the wooden flooring, and, both as regards

Fig. 190.—Portion of north side of Crannog, with the space between inner and
second circle of piles dug out, showing arrangement of Transverse Beams
and structure of the Log Pavement. (*From Photograph by Mr. Laurie.*)

distance and shape, looked as if they had been mortised into
the holes in the former. When the beam was thus applied
and restored into its natural position, the portion of its under
side between the mortised holes was observed to have a
longitudinal groove, having its inner margin bevelled off, and
containing a few round holes, which, however, did not penetrate to its upper surface, and just underneath it were the
external ends of two large oak planks, which extended
inwards to the doorway. On careful inspection these planks

were also found to contain a few vertical holes, so that it
became apparent that the interval between them and the
large transverse was protected by a series of upright wooden
spars. External to this parapet-like arrangement was the
refuse-heap, which, on being entirely cleared away, showed
that the two uprights, though exposed to a depth of about
6 feet below the wooden pavement, were immovably fixed.
Close to one of them deeper digging was attempted, with
the view of getting an idea of its length, and at a depth of
4 feet still lower a solid beam could be felt with an iron
probe; but whether the upright was mortised into it I could
not determine. Continuous with the east end of this ashpit railing was the external circle of stockades, which curved
a little outwards, and at the other end, in addition to an
external line of slender stockades which took a more rapid
sweep outwards, there was a straight row of uprights thickly
placed together, and protected at their base by a strong fixed
beam, into which they were mortised (see Fig. 188). This
beam was on a lower level than the platform in front of the
doorway, and the upper ends of the uprights were free, but
the probability is that originally they were bound by a transverse rail. On the inner side of this line a number of short
beams were observed lying flat, as if they had been intended
for a pathway, and towards its external end there lay a con-
fused heap of slender beams projecting beyond the line of
the outer stockades. It was this peculiarity that suggested
this spot as the probable terminus of an under-water gangway leading to the shore, the determination of which led to
the making of a trench some 12 feet farther out, which
resulted in the discovery of a canoe.

Though nothing in the arrangement of the wooden structures here could be construed to indicate a regular landing-
stage, it was very probable, from its southern exposure, the
position of the canoe, and the proximity of the doorway to

the log pavement, together with the pathway leading up to it, that this really was the ordinary landing-place as well as the outer entrance to the crannog.

Refuse-heap.—As mentioned above, the refuse-heap lay outside the stockades, and immediately beyond the railing in front of the supposed doorway to the central area of the crannog. It was of an oblong shape, measuring from 25 to 30 feet long (along the margin of the island), and about 15 to 20 feet across. Its depth, near the railing, would be about 5 feet in addition to its superficial layer of clay and silt. The principal ingredients of its central portion were broken bones and ashes, but towards the margin and lower strata these were largely mixed with decayed brushwood. To clear out its deeper portions was a difficult matter, owing to the rapid accumulation of water. One of the combs (Fig. 218), and a bone pin, were found here in my presence, at a depth of not less than 6 or 7 feet below the surface of the field. The lowest stratum reached consisted of what seemed to me to be lake silt, brushwood, and some large bones. The bones, especially those from the lower strata, were abundantly impregnated with the mineral vivianite, which, in some of the larger ones, formed groups of most beautiful green crystals, similar in all respects to those found at the Lochlee crannog. What, however, made the investigation of the midden so full of interest was the number of rare and valuable relics recovered from its contents. Some of them were picked up *in situ*, when the men were wheeling out the stuff, but others were subsequently found by riddling the débris when it became sufficiently dry to admit of this process.

The general results of the above observations may be categorically summed up as follows:—

1. The island, as far as could be ascertained from the investigations made, was composed of a succession of layers of

the trunks and branches of trees, intermingled in some places with stones, turf, etc.

2. The whole mass was kept firmly together by a peculiar arrangement of upright and horizontal beams, forming a united series of circular stockades.

3. The outer circle was intended more for protection than for giving stability to the island, and in some parts, as at the east side of refuse-heap, was neatly constructed after the manner of a stair-railing, while the inner one not only gave stability to the island, but was used as a fence, or in connection with the superstructural buildings.

4. The central portion was rudely paved with wooden beams, many of which were firmly fixed to the lower woodwork by stout wooden pegs as well as to the encircling stockades, thus affording here and there, as it were, *points d'appui.*

5. While there was one general fireplace situated near the centre, evidence of occasional fires elsewhere was quite conclusive, one of which appeared to have been a smelting-furnace.

6. The entrance to the central area was looking south-east, and in front of it there was a well-constructed wooden platform, made of large oak planks, supported on solid layers of wood to which they were pinned down.

7. Beyond the platform, but separated from it by a massive wooden railing, was the refuse-heap ; and to the right of it a pathway, also protected on its outer side by a railing, led downwards and westwards to the line of the outer circle, where there appeared to have been an opening towards a rude landing-stage at the water edge.

8. As to the kind of dwelling-house that no doubt once occupied this site, whether one large pagoda-like building or a series of small huts, the evidence is inconclusive, but so far as it goes it appears to me to be indicative of the former.

In addition to what has already been stated, there remains to notice only a few broken pieces of wood containing round holes, together with a variety of large and small pins similar to those described and figured in my notice of the Lochlee crannog.

Discovery and Description of Canoe.—The experience derived from the investigations of the crannogs at Lochlee and Lochspouts, in both of which a submerged gangway was found running to the nearest shore, was sufficiently suggestive to keep me on the *qui vive* for any indications of a similar structure here. On the north side, where the shore was nearest, though the digging was carried considerably deeper and farther out from the margin of the crannog than elsewhere, not the slightest appearance of outlying wood-work was observed ; and as there was no probability of an approach from the more distant ends of the lake, the situation of a gangway, if such existed at all, was limited to the south-west side, where the shore would be about 150 yards distant. To determine this, the men were set to cut a trench about 12 yards distant from the crannog, across the most likely line, so as to intercept it, and after going down 4 feet they came upon a layer of brushwood, along with one or two beams, below which there seemed to be the usual lake mud. Upon forcing the spade downwards, however, a hard beam was encountered, which at first I took to be the discovery of part of the gangway we were in search of, and to satisfy myself on the point I took an iron rod, and, by carefully probing all over the bottom of the trench, ascertained that instead of a gangway we had come upon portion of a canoe. Guided by the direction of the supposed side of the canoe, which looked like a thin oak beam running along the edge of the trench, a suitable clearance was made, which revealed to the wondering gaze of the bystanders the front half of a large canoe. Upon being subsequently exposed in its entirety it was

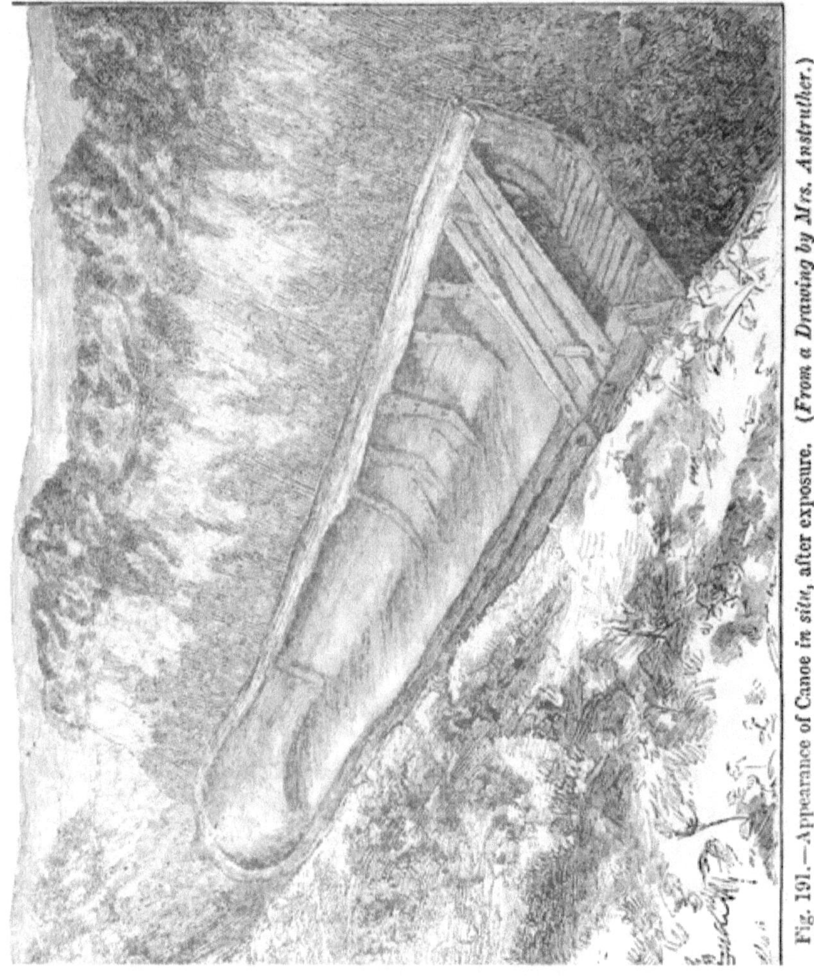

Fig. 191.—Appearance of Canoe *in situ*, after exposure. (*From a Drawing by Mrs. Anstruther.*)

found to have the following dimensions and peculiarities. Its sides were supported by a series of well-shaped ribs, which extended from the rim to near the mesial line, and sometimes a little beyond it. This, at first sight, gave the canoe the appearance of a boat, but after careful inspection it became apparent that these ribs were no part of the original vessel, but subsequent additions made for repairing and strengthening purposes. Nearly the whole of one side was lined with broad thin boards made of soft wood, external to which was the thin oak side of the canoe, having its cracks as well as the intermediate spaces between it and the strengthening boards actually stuffed with a species of moss. Moreover, the ribs on this side were more numerous than on the other side, no less than ten having been observed on the former, and their arrangement on both sides was totally devoid of regularity. Of the whole series of ribs only two were made of oak, the rest being of birch or some perishable wood, and so decayed that it was with great care they were prevented from being entirely destroyed by the workmen, as they offered no resistance to their spades. They were fastened to the canoe by wooden pins, arranged generally in couples forming two rows along the rib, and so closely were they placed that not less than sixteen were counted in one rib. In two places the canoe had been repaired by inserting a nicely fitting piece of oak planking instead of the original portion of the side. One of these patches measured 2 feet 3 inches long, and 10 inches broad, and was kept in position by two ribs, one near each end. The stem, which was symmetrically shaped and pointed, was pierced horizontally by a large hole, and about 3 feet from its tip each side had also an elongated hole near the rim, sufficiently large to admit of being easily grasped by the hand. Externally, and on both sides, there was fastened to the edge of the canoe, by means of wooden pins, a sort of gunwale, which extended from within a few feet of the stem till it projected a little beyond the stern. Close to the stern, two slender bars of wood, a few inches apart, stretched across, and after passing through the edges of the canoe terminated in being tightly mortised into the gunwale. These transverses contained two round holes similarly arranged as to position, being near the right side, and between them was inserted a moveable sternpiece which was shaped to the curve of the canoe, *i.e.* approximately a semicircle, and made to fit into a shallow groove cut out of the solid wood. This sternpiece was strongly constructed, being $3\frac{1}{2}$ inches thick, 3 feet 6 inches long, and 1 foot $4\frac{1}{2}$ inches deep about the middle. About 15 inches in advance of the sternpiece there was a ridge across the bottom and sides of the canoe which looked like a rib, but was really part of the solid oak, evidently left for a special purpose. I also noticed one or two round holes in the floor, as well as others along its upper edge as if for thole-pins. In two places equidistant from the ends, and about 4 feet apart, the gunwale had short pieces of wood fastened to it by vertical pins, as if intended for the use of oars. Amongst the decayed brushwood lying across the canoe was an oak beam, having one end projecting so much beyond the edge into the clay bank that the workmen in endeavouring to pull it out broke off the free end. This portion was rectangularly shaped, 5 inches by $3\frac{1}{2}$, and had its narrow side pierced with three round holes 1 foot 10 inches apart, which still contained the remains of broken pins. The shell of the canoe was oak, made by scooping out the interior of a large trunk, but all its attachments, such as gunwale, sternpiece, cross spars at stern, and all the ribs except two, were made of a much less durable wood.

The extreme length of the canoe was 22 feet, but the inside measurements were as follows:—Length 19 feet

6 inches; breadth at stern 3 feet 6 inches; ditto, about the middle, 4 feet; and ditto, near the stem, 2 feet 10 inches; depth, about centre, 1 foot 10 inches.

Among the mud removed from the hull of the canoe were a few stones and portion of the skull of an ox (see Fig. 191).

Oar.—Portion of what appeared to have been a large oar was found on the crannog, but, from its fragmentary state, we could only ascertain that the blade was 9 inches broad and $1\frac{1}{4}$ inch thick, and that the handle measured 5 inches in circumference.

DESCRIPTION OF RELICS.

The relics are here grouped under several heads, in accordance with the method of classification adopted in my previous monographs, and, to save repetition, I may explain, that (when not otherwise stated) they may be considered to have been found either *in situ* in the refuse-heap, or among its stuff after it was wheeled out and subsequently examined.

I. OBJECTS MADE OF STONE.

Hammer-Stones, Polishers, etc.—Only two or three typical hammer-stones have to be recorded as found on this crannog.

Fig. 192.—Stone Polisher (⅔).

One is an elongated flat pebble, and shows the usual markings at both ends, another only at one extremity, and a third is somewhat circular, with the markings on the flat surface alone. Under the category of polishers are included seven or eight highly polished water-worn pebbles, varying much in size and shape. Two, shaped like pebbles, are 7 inches long, and have slight pounding marks at both extremities (Fig. 192). Three are flat and oblong, and measure from $2\frac{1}{2}$ to 4 inches.

Sling-Stones, etc.—Like the hammer-stones, these objects are comparatively rare, only a few having been added to the collection.

Whetstones, Grindstones, etc.—Of these objects the following are noteworthy :—

1. A large flat implement, made of bluish claystone, with a smooth polished surface, and having a hole roughly cut out of one end. It measures 12 inches long, 4 broad, and $1\frac{1}{2}$ inch thick (Fig. 193).

Fig. 193.— (?) Whetstone ($\frac{1}{3}$).

2. One or two ordinary whetstones a few inches long, and from 1 to 2 inches broad.

3. An oblong block of sandstone, containing two smooth cavities, probably used for polishing small objects such as jet rings. One of the cavities is a hollowed circle $2\frac{1}{2}$ inches

in diameter, and half an inch deep; the other is a groove
3 inches long, half an inch wide, and the same in depth
(Fig. 194).

4. Two fragments of a circular grindstone, made of fine
red sandstone. One of the portions shows a few inches of the striated circumference as well as a small segment of the central hole. The diameter of the stone when whole would be about 15 inches.

Fig. 194.—Block of Sandstone (¼).

5. Two large irregularly-shaped masses of whitish sandstone, each containing a smooth cavity shaped like a trowel or botanical spud, having the sides curled up.
One of these curiously-shaped cavities measures 10 by 8
inches. Its greatest depth, which is at the base and in
the line of the shortest diameter, is 3 inches. The other
is precisely similar in shape, but of smaller dimensions.
The latter stone has friction-marks on another of its
sides.

6. Another mass of whitish sandstone, of a semi-globular
shape, having a cup-shaped cavity on its flat surface, must
also be included under this heading. The diameter of the
cup is 5½ inches, and its depth 2½ inches. The rest of the
flat surface all round the margin of the cup is smoothed and
striated, evidently caused by the sharpening of tools. The
cup itself was not used for this purpose, as the marks of the
puncheon by which it was chiselled out are distinctly seen.
Its probable use was to hold water, so essential to the
sharpening of metal tools.

Cup Stone.—A small cup stone found in the interior of the crannog. The stone is smooth on its upper and under surfaces and on one side, but the other sides are irregularly shaped. The cup itself is quite smooth and circular, and looks as if it had been used as a small mortar. Its diameter is only 1 inch, and depth half an inch (Fig. 195).

Querns.—Only two upper quern-stones, both of which are in a fragmentary condition. One was made of a fine

Fig. 195. Fragment of Stone, with a cup-shaped Cavity (⅓).

quartz conglomerate, and, by putting the fragments together, it was ascertained that it measured 18 inches by 17 inches. It was flat, and more of a millstone shape, and the central hole was large (3 inches in diameter), circular, and not tapering. For the insertion of a handle there was a small square-shaped hole at its margin.

Portion of another quern made of whinstone, and of the usual type, indicates a medium size, of about 1 foot across.

Fig. 196.—Spindle Whorl (⅓).

Fig. 197.—Spindle Whorl (⅓).

Spindle Whorls.—A small spindle whorl neatly made of coarse shale. It is flat and circular, and has a diameter of 1 inch (Fig. 196). Another perforated little object, of smaller dimensions than the former, is made of cannel coal (Fig. 197).

Flint Objects.—Two views of a portion of a curved flint knife, which has been much used, are here given (Fig. 198). Another small flint implement like a scraper is figured, because it exhibits one side which has been artificially polished (Fig. 199). Fig. 200 represents a small central core, neatly chipped all round. There is another large core of flint 3¼ inches in diameter, from which many flakes appear to have been struck off. Besides the above there were found a small portion of a finely chipped scraper, and a large quantity of broken flints and chips.

Fig. 199.—Polished Flint Implement (½).

Fig. 198.—Flint Implement (⅓).

Fig. 200.—Flint Core (⅓).

Finally, small pebbles, sometimes highly polished and variegated in colour, thin circular discs of stone about the size of a halfpenny-piece, bits of dark shale as if water-worn, and a large quartz crystal having its angles worn off, may be mentioned among the nondescript articles under this heading. Also a lump of iron slag was found near the middle of the first trench, but, mysteriously, it could not be seen when collecting the objects at the end of the day's work, and was never recovered.

II. OBJECTS OF BONE.

Pins.—Twenty bone pins, varying in length from 1½ inch to 3½ inches. These articles are exceedingly well made, with round polished stems, tapering into sharp points. Some have round heads like beads, others are circular but

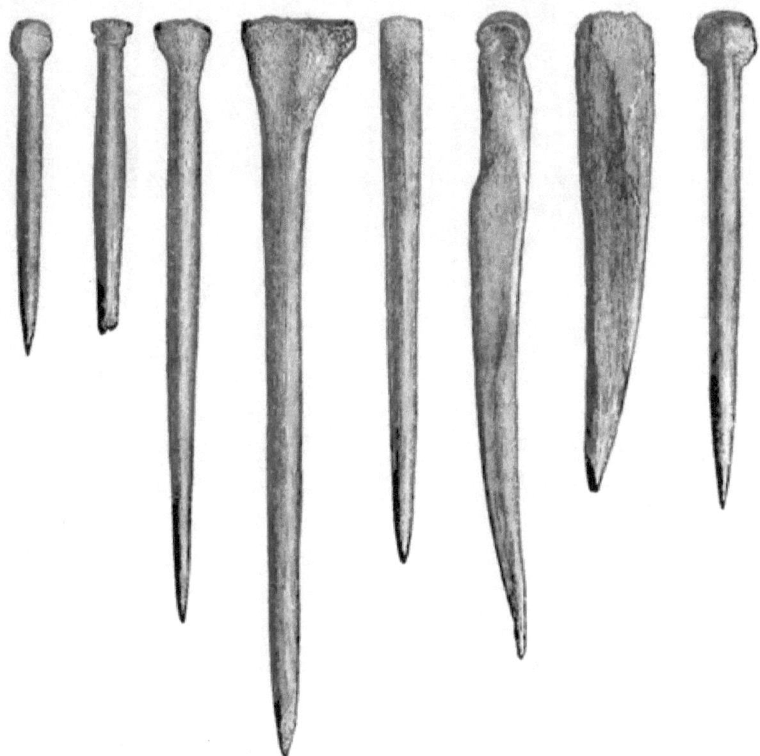

Fig. 201. Fig. 202. Fig. 203. Fig. 204. Fig. 205. Fig. 206. Fig. 207. Fig. 208.
Bone Pins (⅓).

flat on the top, while others again, especially the larger ones, are irregularly shaped. One (Fig. 202) has its head ornamented by a circular ridge, surmounted by a wider rim neatly notched all round, and another has its shank surrounded by two bands of diamond-shaped spaces, formed by

a series of incised lines slantingly crossing each other, as
shown in Fig. 212. Fig. 210 is the representation of one
only partially formed.

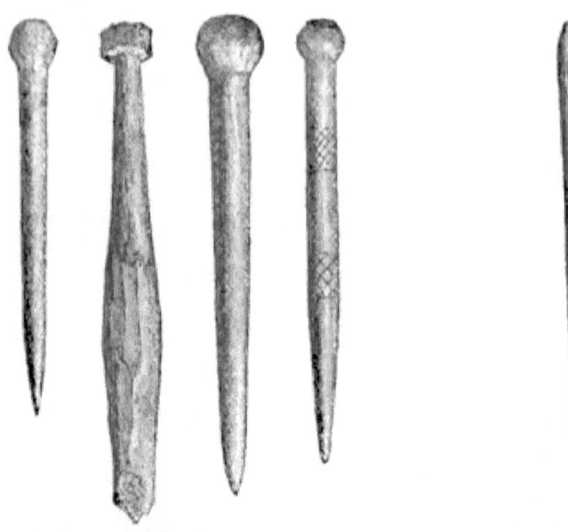

Fig. 209. Fig. 210. Fig. 211. Fig. 212.   Fig. 213.
Bone Pins (⅓).   Bone Needle (⅓).

Needle.—A neatly-formed needle, having an elongated
eye at its extremity, precisely similar to a common darning
needle. It tapers gently into a sharp tip, and is smoothly
polished all over. Its length is 2 inches (Fig. 213).

Fig. 214.—Bone Knob (¼).

Fig. 215.—Bone Knob (⅓).

Knobs.—Three round objects of bone, about the size of a
marble, each having a portion of a slender iron pin more or
less projecting. Two are quite smooth, globular, and precisely

similar to each other in every respect (Fig. 214); the other is ornamented by a few incised circles and ridges (Fig. 215).

Fig. 216 represents a curiously-shaped object of bone, the use of which is unknown to me.

Worked Bones.—Several portions of bone, exhibiting marks of sharp-cutting instruments, but not assuming the form of any recognisable implements.

Toilet Combs.—Three of these articles, which are in a wonderfully good state of preservation, are here engraved on account of their structure and variety of ornamentation.

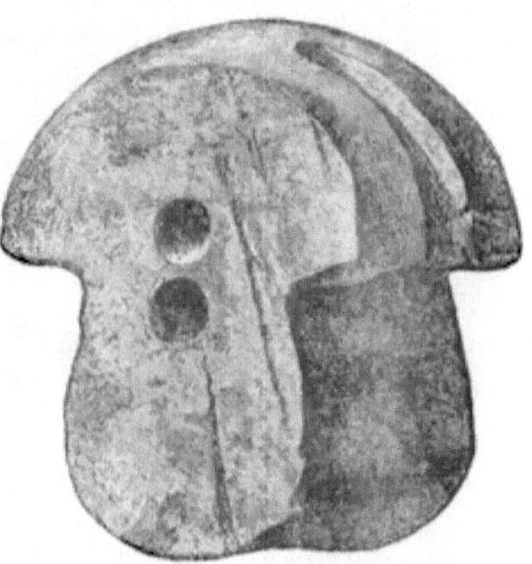

Fig. 216.—Object of Bone (¼).

They are all made on a uniform plan. The body, *i.e.* the portion containing the teeth, consists of three or four flat pieces kept in position by two transverse bands of bone, one

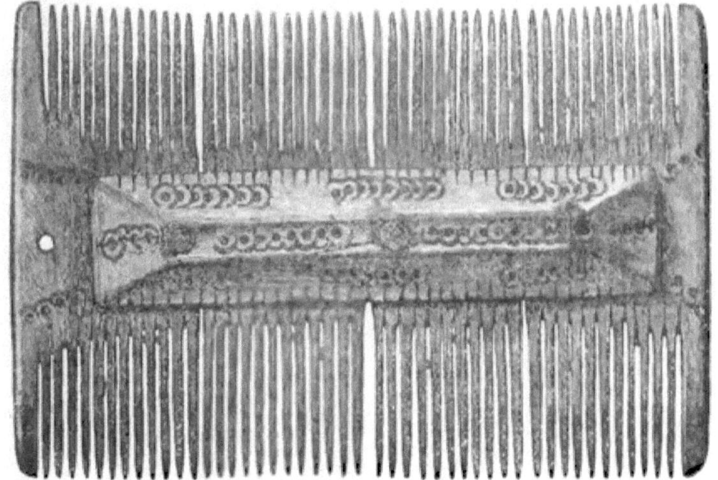

Fig. 217.—Bone Comb (½).

on each side, and riveted together by three or four iron rivets.
The comb represented by Fig. 217 has its body made of four

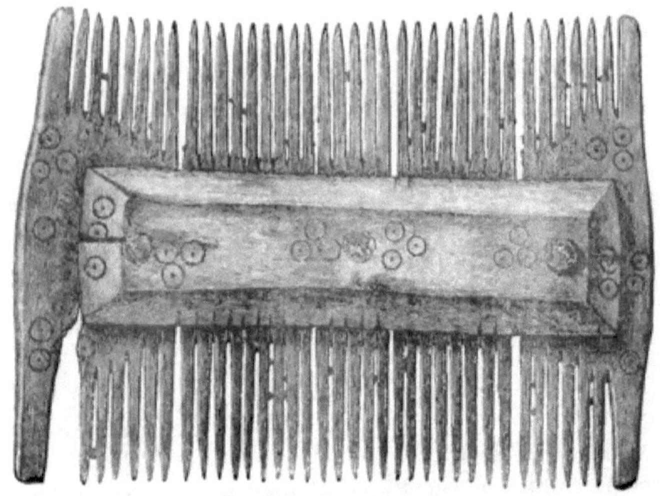

Fig. 218.— Bone Comb (⅓).

portions, but only three rivets.   The ornamentation is alike
on both sides, and at one end there is a small hole, probably

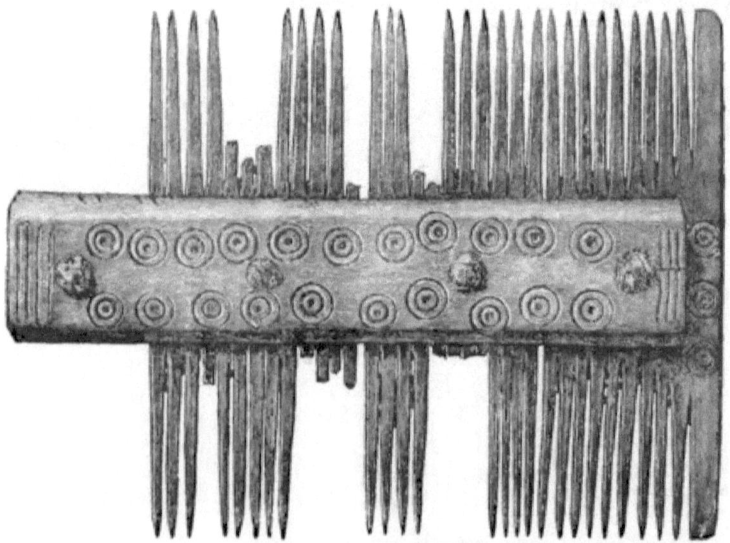

Fig. 219.—Bone Comb (⅓).

for attaching it to a string.   It is 3½ inches long and 2¼ inches broad.   That figured next (Fig. 218) has the same breadth as the former, but not quite the same length, being only 3 inches long.   The ornamentation is similar on both sides.

From slight cuts on the cross bars, corresponding to the intervals between the teeth, it is manifest that the teeth in both these combs were formed by a saw, after the pieces were riveted together.

The third comb here engraved (Fig. 219) is in a somewhat fragmentary condition, but when whole it would be about 4 inches long.   The body was made up of four portions, and contained four iron rivets.   Its ornamentation consists of a central dot, surrounded by two incised circles, and is alike on both sides.   The similarity of the concentric circles induces me to believe that they must have been formed by a die, probably branded on with a hot iron.

Some other fragments of similar combs were found, representing at least three additional combs, with teeth rather finer than those in the illustrations.

III. Objects made of Horn.

Several portions of deer horns, consisting of tines and thick portions of the body of the horn, together with a few of the roe-deer, presenting sometimes marks of a saw and sometimes those of a sharp-cutting tool, were found in the refuse-heap.   The few worked objects I have to record were all made from horns of the former animal.   One large antler, having portion of the skull attached to it, and the entire lateral half of the skull of a roe-deer with the horn still adherent, show that the horns were not shed ones, but those of animals actually caught and killed.   The manufactured implements consist of a few pointed objects, and one or two handles, apparently for knives.

Fig. 220 represents a highly polished dagger-like implement, measuring 7½ inches long. Another, of about the same size, is coarsely cut out of the side of a large horn (Fig. 221). A small pointed object is figured among the bone pins (see Fig. 207).

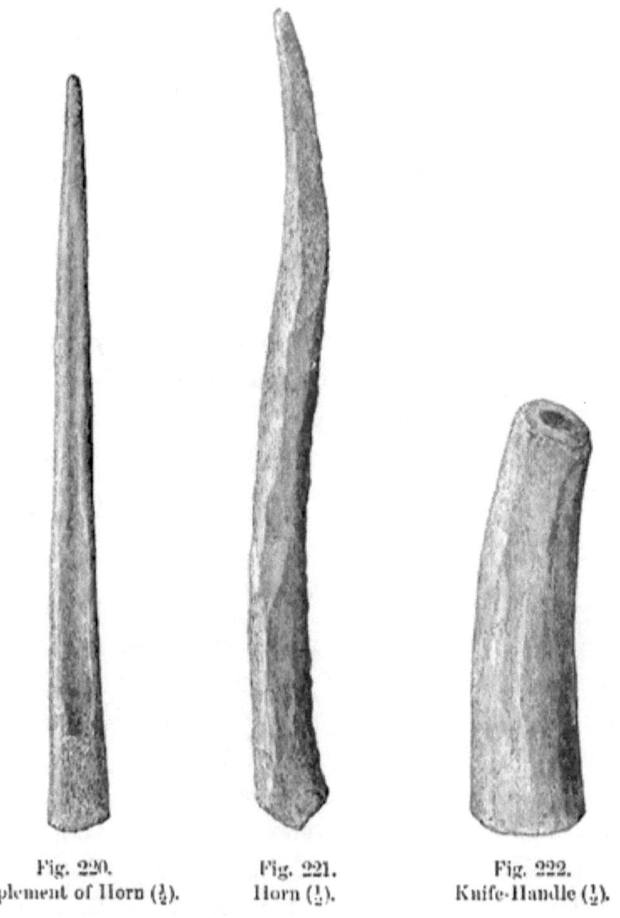

Fig. 220.
Implement of Horn (½).

Fig. 221.
Horn (½).

Fig. 222.
Knife-Handle (½).

Knife-Handles.—One of the handles is well made, having the rough surface removed with a sharp-cutting instrument. It is 4 inches long (Fig. 222). Another is only 3 inches long, and has a notch at one end.

IV. Objects of Wood.

Wooden objects are extremely rare. One or two fragments of what appeared to have been a bowl, portion of the blade of an oar, a bit of board partially burnt and penetrated by four round holes, together with three pins almost identical with those found at Lochlee (see Figs. 112, 114, and 115). The bowl was ornamented by two or three incised parallel lines near the rim. Another small fragment, which might have been of the same vessel, had a clasp of thin brass over it, as if it had been mended.

V. Objects of Metal.

(a.) *Articles made of Iron.*

1. *Axe Head.*—This implement, which is represented in Fig. 223, measures 3 inches along the cutting edge, 4½ inches

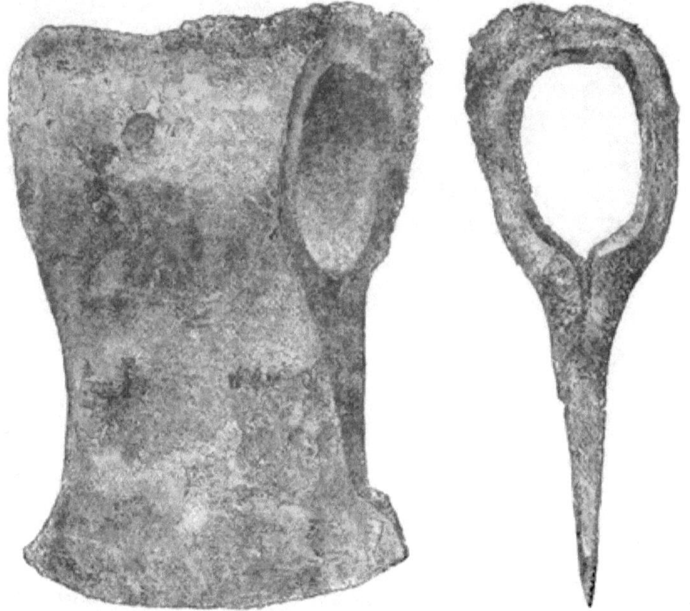

Fig. 223. Iron Axe (⅓).

from the centre of cutting edge to the back of the hole for
handle, and 2 inches through the centre of this aperture.   A
neighbouring farmer, who had carted a load of the stuff from
the midden for potting plants, found this axe-head while making use of the stuff in his greenhouse, and returned it to me.

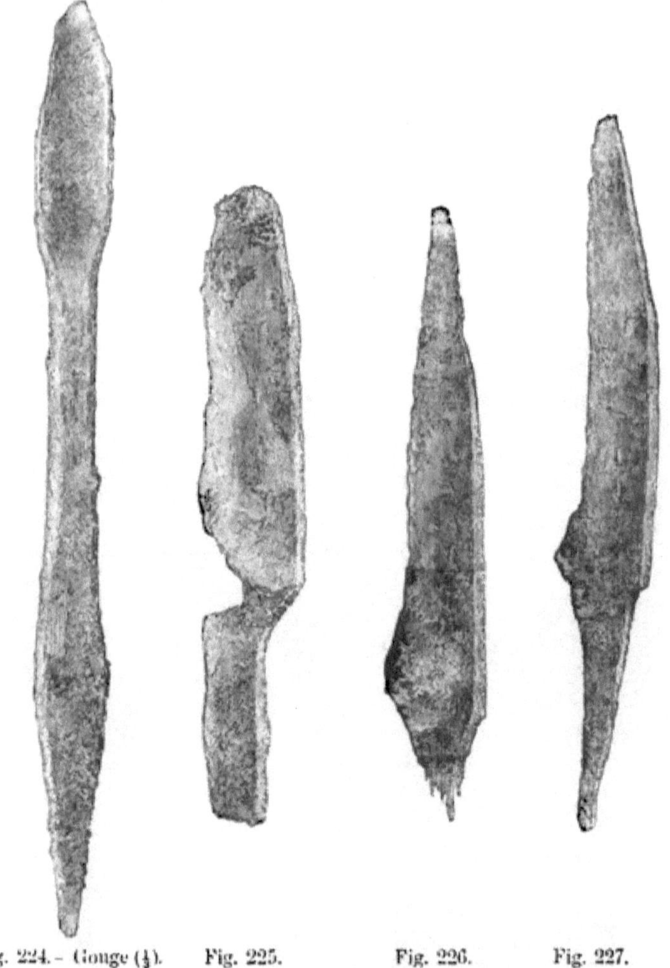

Fig. 224.- Gouge (⅓).   Fig. 225.   Fig. 226.   Fig. 227.
Iron Knives (⅓).

2. *Gouge.*—This instrument appears to have had a portion broken off its point.   It still measures 14 inches long,

and its other extremity is pointed for insertion into a handle
(Fig. 224).

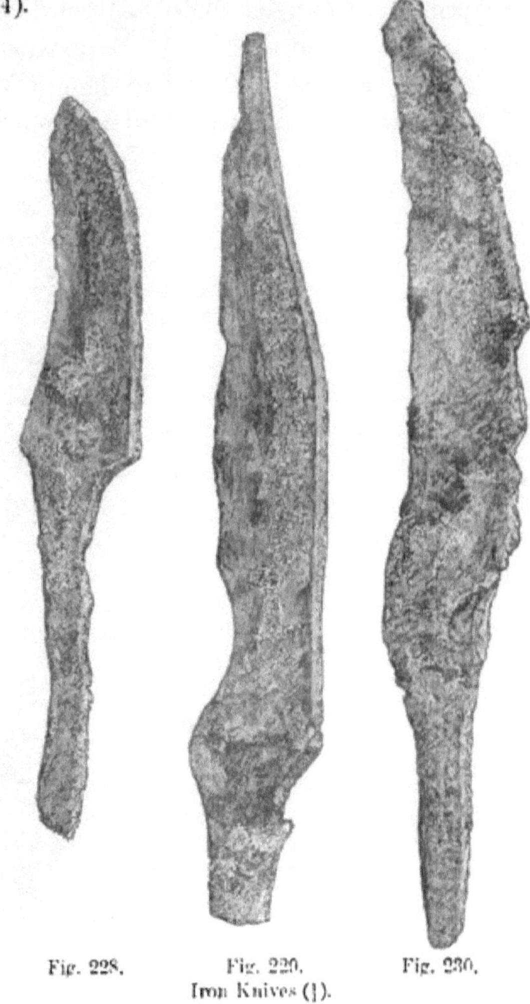

Fig. 228.   Fig. 229.   Fig. 230.
Iron Knives (⅓).

3. *Knives.*[1]—Six well-shaped knife-blades, all with tangs
for insertion into handles. The blades vary in length from
2 to 4 inches (Figs. 225 to 230).

[1] In the Museum of the Royal Irish Academy I noticed several knives
precisely similar to those here figured, which were found on the crannog
of Ballinderry.

4. *Punch.*—This implement is 6 inches long, and rectangularly shaped, with its angles slightly flattened (Fig. 231).

5. *Awls.*—Of these objects there are three: one is very slender and sharp, but only 2 inches long (Fig. 232). Another is 4 inches long, and the third is a much larger implement, being 7¼ inches long.

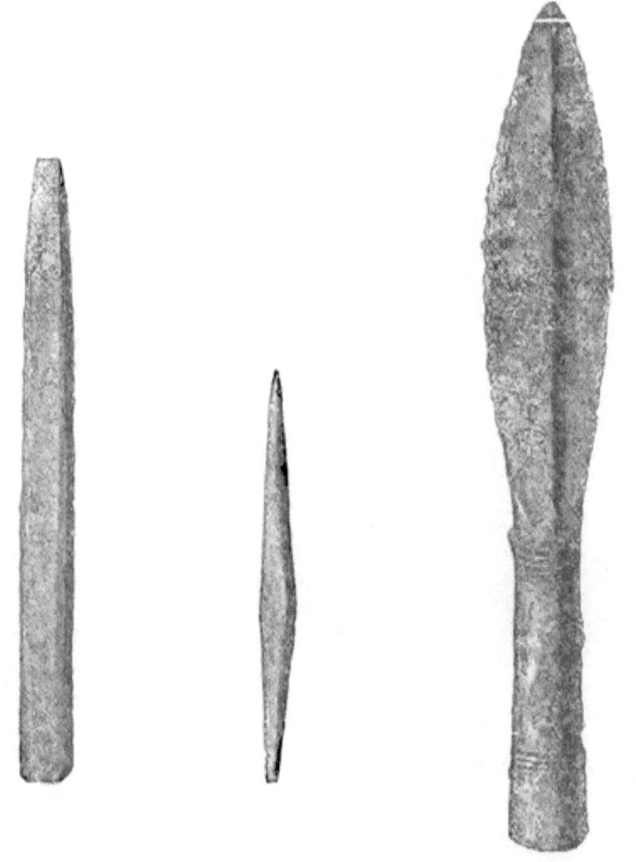

Fig. 231.—Punch (½).   Fig. 232.—Awl (½).   Fig. 233.—Spear Head (½).

6. *Spear Head.*—This is a well-shaped socketed spear head, 8½ inches long, with a central ridge in the blade. The socket end is ornamented by two groups of circular grooves,

each group containing three circles.   Portion of the wooden
handle was found in the socket (Fig. 233).

7. *Arrow Heads.*—Three pointed objects like arrow heads
are represented in Figs. 234, 235, and 236.   Two of these
objects are almost identical in size and form.   One end is
four-sided and tapers to a sharp point, the other is round and

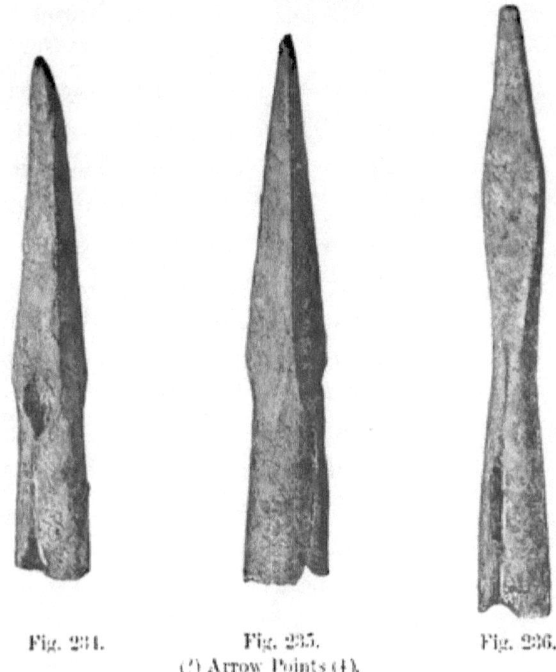

Fig. 234.   Fig. 235.   Fig. 236.
(⁷) Arrow Points (½).

hollow as if for the insertion of the stem of an arrow.
Length 2¾ inches.   The third has the socket end very similar
to the former, but the front portion is flat, and widens out a
little before coming to a sharp point (Fig. 236).

8. Fig. 237 represents a curious object, having a spring
attached to each side, both of which are still compressible,
and a curved portion containing a round hole.   Total length
is 5 inches, length of springs 2 inches, length of curved
portion 1¾ inches.   Said to be portion of a padlock, similar

in structure and principle to locks now used in China and some parts of India.[1]

9. *Files?*—An object shaped like a flat file, cut squarely

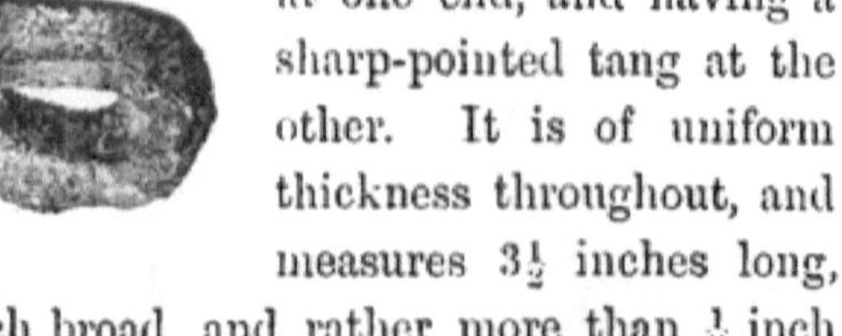

at one end, and having a sharp-pointed tang at the other. It is of uniform thickness throughout, and measures $3\frac{1}{2}$ inches long, $\frac{5}{8}$ inch broad, and rather more than $\frac{1}{8}$ inch thick. There is another object exactly similar to the above in form, but a shade smaller. They look like small files, but no grooves now remain.

10. *Spiral Objects.*—Fig. 238 represents a slender iron rod, forming a close spiral with three twists at one end, and a slight curve at the other, which presents the appearance of having been fractured. The diameter of the circular portion is rather less than 1 inch. Fig. 239 represents another spiral object terminating in a straight point.

11. Fig. 240 represents two views of a small ornamental instrument with a bifurcated termination, which might have been used as a compass for describing small circles, such as are seen on some of the combs. Its length is 2 inches.

12. *Miscellaneous Objects.*—When the stuff wheeled from the refuse-heap had dried up and become pulverised during the summer months, several articles were picked up by visitors, among which may be mentioned four large nails, a small ferrule, a small

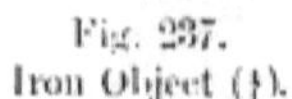

Fig. 237.
Iron Object ($\frac{1}{1}$).

<hr>

iron link thicker on one side than another, a much-corroded
socket still containing a bit of wood, a flat portion of iron

Fig. 238.—Spiral Object (⅓).   Fig. 239.—Iron Object (⅓).

welded together, and a few other bits of iron. These,
however, cannot be positively asserted as belonging to the
crannog objects.

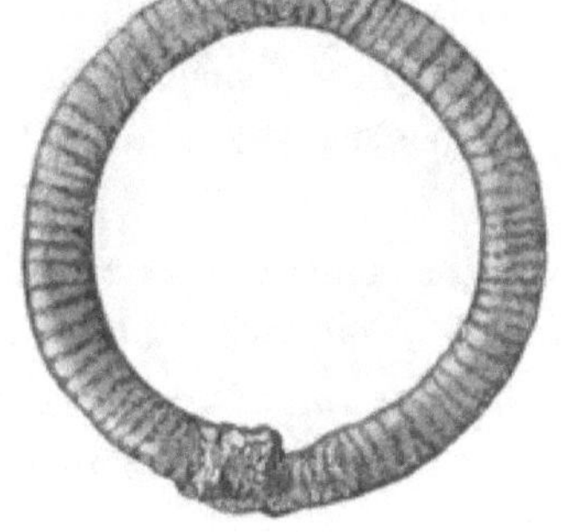

Fig. 240.—Iron Object (⅓).   Two Views.   Fig. 241.—Bronze Brooch (⅓).

barbed locks, with special reference to their presence in crannogs, which
will be read at an early meeting of the Society of Antiquaries of Scotland.
Meantime he writes as follows :—" Padlocks with barbed bolts are almost
the only kind used all over China and India at the present day, and it is
only reasonable to suppose that they are of Eastern origin. They have
been found in England in connection with Roman remains (see paper on
Locks found at Great Chesterfield, Essex, by the Hon. R. Cornwallis
Neville, *Archæolog. Journal*, vol. xiii. p. 7) ; and I think it probable they
were introduced into this country by the Romans. Their use in this
country continued to mediæval times, after which they disappeared before
improved locks of more modern construction."

(b.) *Articles made of Bronze.*

Brooch.—A circular brooch, minus the pin, 1½ inch in diameter, and ornamented on its upper surface by a series of grooves pointing to the centre of the brooch. The under surface is quite plain. A small portion of the pin is still attached to the brooch, and the opposite side of the brooch is worn into a hollow by the friction of the point of the pin. The transverse grooves are also much worn, but where nearly obliterated the external and internal margins of the brooch show the hacks, corresponding with their extremities (Fig. 241).

Pins.—Two small pins, having round shanks ornamented by two groups of circular and longitudinal incised lines. Both pins have flat heads, and one has a blue bead stuck in its top. They are nearly of the same length, being a shade less than a couple of inches (Figs. 242 and 243).

Several bits of brass plate, apparently used as clasps for mending purposes. One, indeed, was found attached to a small portion of a wooden bowl. Also a thin brass button 1¼ inch in diameter.

Fig. 242.　　Fig. 243.
Bronze Pins (⅓).

(c.) *Articles made of Gold.*

Finger-Rings.—On the 14th December one of the workmen while clearing out the refuse-heap turned up a curious spectacle-like ornament, made by twisting the ends of a thick and somewhat square-shaped gold wire into the form of a double spiral ring (Fig. 244). Upon close inspection it became evident that originally this article was a handsome spiral finger-ring, containing 5½ twists, but that, from some means or other, two of the twists had been forced apart from the

others. The direction of certain scratches, and a slight
mark as if a blow had been struck (probably the spade of
the finder), seem to me to confirm this explanation. It lay
buried half-way down in the midden, close to the base of
the large parapet in front of the entrance to the area of the
log-pavement. It weighs 300 grains, and its internal diameter measures a shade over $\frac{5}{8}$ of an inch. On the 16th
April, while clearing away the soil on the west side of the
crannog, a few feet to the inner side of the inner circle of
piles, another spiral ring was found (Fig. 245). It is made

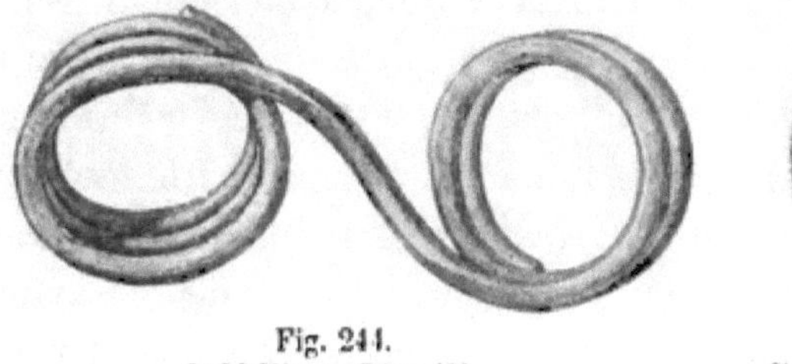

Fig. 244.
Gold Finger-Ring (⅔).

Fig. 245.
Gold Finger-Ring (¼).

of round gold wire, not quite so massive as the former, and
contains rather more than six twists. Both ends taper
slightly, and, for nearly a whole twist, are ornamented by a
series of circular grooves which gives them some resemblance
to the tail end of a serpent. The colour of the gold of this
ring is a brighter yellow than that of the former. Its internal
diameter is exactly $\frac{3}{4}$ of an inch, and its weight is 245 grains.
Both rings were quite clean, and free from all tarnish when
exposed.[1]

[1] In *Prehistoric Annals of Scotland*, vol. i. p. 474, is figured a spiral
bronze ring, of three twists, found during the construction of a new road
leading from Granton Pier to Edinburgh, in a small stone cist, distant
only about twenty yards from the seashore, which has called forth the
following remarks from its learned author : " Examples of the spiral
finger-ring have been repeatedly found in Britain with remains of different periods. They are also known to northern antiquaries among the
older relics of Denmark and Sweden. This may indeed be regarded as
among the earliest forms of the ring, since it is only at a comparatively

Coin.—Mr. Robert Dunlop, iron-moulder, a native of Kilmarnock, but now residing at Airdrie, happened to visit his friends at the beginning of the year, and hearing of the discoveries at the Buston crannog, took the opportunity of visiting it. It was not, however, idle curiosity that prompted him, but a true spirit of inquiry, which often ere now led him to wander abroad as a humble student of nature, and on one occasion even as far as the famous Kent's Cavern. Being a Science teacher in Chemistry he was desirous of securing specimens of the different forms of vivianite, and so picked up from amidst a mass of broken bones and ashes that had just been wheeled from the midden, a lump of a bluish pasty substance, thinking it to be the amorphous form of this mineral. He carried this lump home with him for the purpose of analysing it, but, owing to other duties, was unable to do so till some three months afterwards. Having then taken a portion of the bluish mass, he mixed it with water in a test-tube, and on proceeding to dissolve it, noticed a yellow speck in this blue material. Curious to know what

late period that traces of any knowledge of the art of soldering among native metallurgists became apparent. A silver ring of the same early type formed one of the celebrated Norrie's Law hoard, found on the opposite shore of the Firth of Forth." Dr. Schliemann, in giving an account of the discovery of a treasure in a tomb at Mycenæ, writes as follows : "There were further found four spirals of wire, five plain gold rings, and a similar one of silver, of which a selection is represented under No. 529. I remind the reader that similar spirals and rings of thick gold wire occur in the wall paintings of the Egyptian tombs. They are supposed to have served as presents, or perhaps as a medium of exchange."—(*Mycenæ and Tiryns*, p. 354.) Judging from the paucity of gold spiral finger-rings in our Museums they appear to have been rare. Among the collection of antiquities in the Museum of the Royal Irish Academy, so rich in gold articles, I find only one, thus referred to in Wilde's Catalogue, "a five-sided bar of gold, flat on the inside next the finger, and angular externally, weighing 1 oz. 12 dwt. 6 grs. It may be denominated a torque-ring" (see page 82, fig. 610, Catalogue). In the Belfast collection I also noticed a gold finger-ring with five twists, and having the two ends flattened.

this could be he emptied the tube of its contents, and found
what seemed to be a small gold coin doubled up.   The
slightest effort to restore the coin to its proper shape detached
the portions, and almost at the same moment each portion
separated into two thin plates. Mr. Dunlop then observed
that between the two plates there was a layer of a dark
brittle substance which he most judiciously collected into a
small glass tube for further analysis.   Having then carefully
cleaned the four little plates with a weak solution of nitric
acid, he had the satisfaction, on putting them together, of
restoring the shell of an antique coin, which, as will be seen
from Fig. 246, retains its impressions and characters on both
sides wonderfully distinct. This valuable contribution to
the collection I received at once from its discoverer, as well
as the above narrative of its discovery.

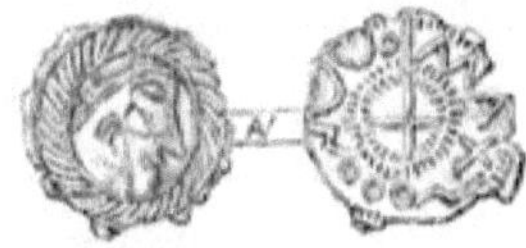

Fig. 246.
Coin found in Buston Crannog.

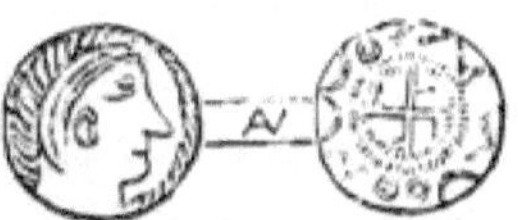

Fig. 247.—For comparison, from
Smith's *Coll.* vol. i. pl. xxii. 9.

Mr. Cochran-Patrick, M.P., to whom I immediately forwarded the different portions of this coin carefully arranged
under a glass slide, as well as the glass tube containing
remains of its core, submitted them to the consideration of
J. Evans, Esq., F.R.S., F.S.A., so well known for his special
knowledge of ancient British coins.

The following interesting remarks by Mr. Evans on the
subject have been sent to me by Mr. Cochran-Patrick :—

" The two plates of gold seem originally to have formed
the shell of an early forgery of a coin, the oxidised core of
which forms the contents of the small tube. I thought at
first that the substance might be resinous, but I think it is

some salt of copper.[1] Some chemist could readily try this.
The coin itself belongs to a class of trientes which have
been found almost exclusively in England, and are probably
of Saxon origin. Enclosed is an impression of one found
near Dover. See Smith's *Coll. Ant.*, vol. i. pl. xxii. 9. Others
were in the Bagshot Heath or Crondale find. See *Num.
Chron.*, N. S., vol. x. 164, pl. xiii. 24 to 26; *Num. Chron.*,
vol. vi. They probably belong to the sixth or seventh century. The find is of value as helping to assign a date to the
crannog." (Figs. 246 and 247.)

VI. MISCELLANEOUS OBJECTS.

1. *Armlets.*—Fragments of three armlets made of cannel
coal, very similar to those found at Lochlee and Lochspouts.

2. *Jet Ornament.*—A small link-shaped ornament of jet,
with two small holes for attachment in one side (Fig. 248).
This object was found on the surface of a mound of débris
long after it was wheeled out, and hence no dependence can
be put on its antiquity.

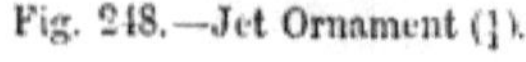

Fig. 248.—Jet Ornament (⅓).   Fig. 249.—Bead (⅓).

3. *Beads, Vitreous Paste, etc.*—A cylindrically-shaped
bead, variegated with three different colours, red and yellow
predominating over patches of transparent glass (Fig. 249).

---

[1] Mr. Dunlop, the finder of the coin, and Mr. John Borland, F.C.S.,
F.R.M.S., Kilmarnock, analysed this substance, and both pronounced it to
be a salt of copper.

Half of a tiny yellow bead, of a vitreous substance, only $\frac{3}{14}$ of an inch in diameter.

A round object, of the size of a small marble, made of vitreous paste, variegated with blue and white, but without any aperture.

Another small flattened object, about the size of a shilling, made of a white compact vitreous substance. It is very smooth, rounded on one side, but flattened on the other. Looks like a drop of a semi-liquid that had fallen on a smooth floor. In the York Museum, case C, amongst some other Roman antiquities I observed several similar articles, which are referred to in the Handbook as "roundlets of coloured glass, probably to set in brooches, from the railway excavations, 1874-75."

One or two little round bits of a dark slag.

4. *Glass.*—Three fragments of thick bright-green glass, all irregularly shaped.

5. *Leather.*—Several strips and chippings of very thin leather.

6. *Pottery.*—A small fragment of Samian ware, only about

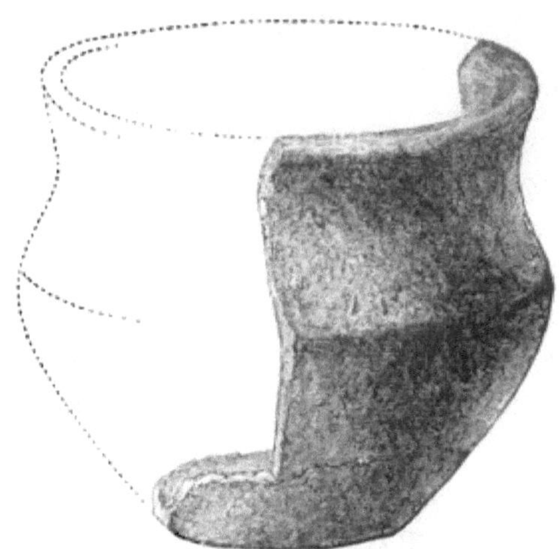

Fig. 250.—Pottery (⅔).

a square inch, with the glaze nearly worn off, but quite un-
mistakeable in its character.

Fig. 250 represents a fragment of a small dish with its
outline. This vessel was made of a hard tinkling ware, black
externally, and of a dull white inside, and measured $3\frac{1}{2}$ inches
across its mouth and 3 inches in depth.

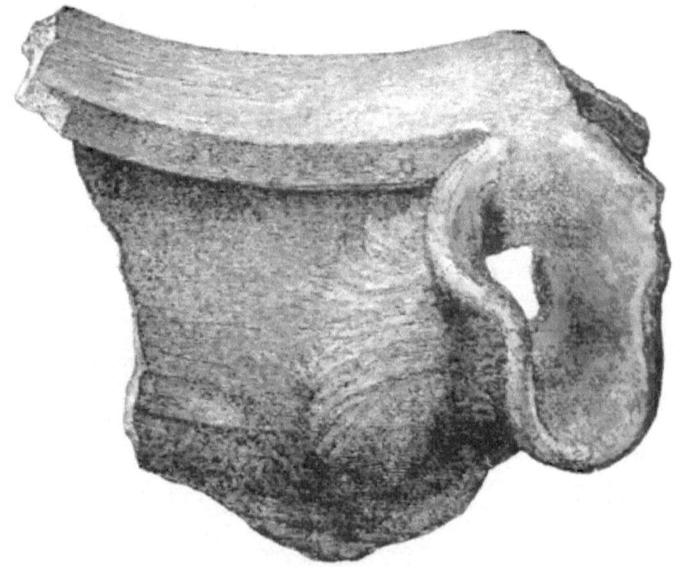

Fig. 251.—Pottery (⅓).

Portion of a large vessel made of coarse materials, having
a short spout just below its everted rim (Fig. 251). The
outside is very black, and the inside has a reddish tinge.
Another portion, apparently of the same vessel, shows the
striation of the potter's wheel.

Fig. 252 represents a curious little knob of pottery.
None of the pottery found here had any appearance of a
glaze.

7. Portion of a small object like a button, made of a soft
chalky substance, is represented in Fig. 253. It shows some
lines as an ornament on its upper surface.

8. *Crucibles.*—A small conical crucible, made of hardened clay arranged in two thin layers, the external of which looks coarser than the other. It has a triangularly-shaped mouth, and at one of its apices there is a slight indentation for facili-

Fig. 252.—Pottery Knob (⅓).

Fig. 253.—(?) Portion of Button (⅓).

tating the pouring out of the smelted material. Its depth is 1½ inch, and circumference of mouth 7 inches. It is cracked all over with heat, and a little dark slag forming a

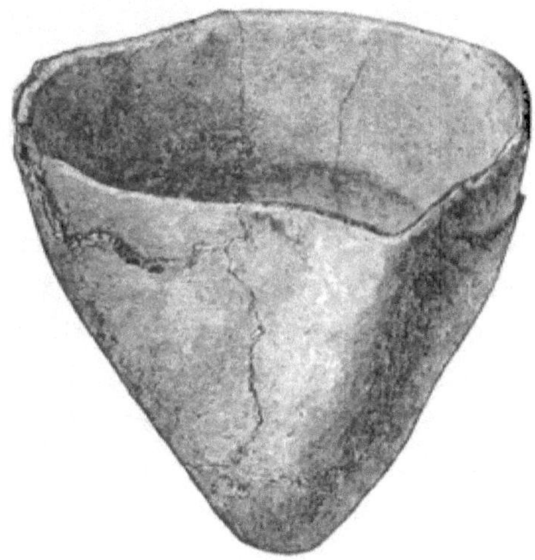

Fig. 254.—Clay Crucible (⅓).

horizontal rim on its inner surface still remains to attest its purpose. This relic was found on the west side of the crannog, not far from the site of the second spiral ring, but outside the inner circle of piles (Fig. 254).

A second crucible, neatly formed and quite whole, was found in the débris wheeled out from the lowest stratum of the refuse-heap. It is of the usual conical form, with a three-cornered mouth about 3 inches in circumference, and measures 1 inch in depth. Particles of a yellowish metal, like brass or bronze, are seen mixed with a kind of slag, near one of the corners. The outside has a glazed appearance, as if it had been subjected to great heat, and to the apex of the cone there is a small bit of cinder still adherent.

Portion of a third crucible, very similar to the last described, was also found at the crannog by a visitor, and publicly exhibited at a bazaar in Kilmarnock.[1] This crucible is interesting as furnishing undoubted evidence that it had been used for melting gold, there being several globules of this metal adhering to its sides, both inside and outside.

REPORT OF OSSEOUS REMAINS FOUND AT BUSTON.

(By Professor Cleland, M.D., F.R.S.)

The osteological specimens obtained from what appears to have been the kitchen-midden of the lake-dwelling at Buston consist in greater part of bones of the ox; while next in frequency are bones of the sheep and the pig. A calcaneum and astragali of the red-deer have been found, as also portions of large red-deer horns, and two portions of roe-deer skull with horns attached. In addition a radius and metacarpal of a goose were found.

The bones of the pig were both full-grown and young; the full-grown, with the teeth worn, being apparently most

---

[1] Along with a few other relics here exhibited (most of which, I believe, were taken from the Buston crannog) were—the bone pin represented by Fig. 212, a small bronze ring, an iron knife-blade, and a fragment of pottery which was found to fit exactly into that represented by Fig. 251.

abundant. They have belonged to an animal of small size, similar probably to that whose remains are found in other Ayrshire deposits.

The remains of the ox and the sheep I account more interesting on account of variety among them.

Ox.—Examining six portions of ox skull, I find one with the horn-core represented by a mere nodule; two specimens each with a portion of horn-core 2·8 inches in greatest diameter, one with a horn-core 2·2 inches diameter at base, and two others with horn-cores 1·8 inches in greatest diameter at base, and one with a horn-core 1½ inch diameter. All the horn-cores are fragmentary; but I judge that none of the last three could have exceeded 5 inches in length, while the first two must have been much longer. Only one of these specimens, that with the smallest horn, has the suture above the occipital bone open. The others must have been adult; and we may judge that we have not to deal with mere aboriginal *Bos longifrons*, but with varieties of ox. The variation seems not to have been confined to the horns. Among a number of first phalanges the majority were slender and small, but there was considerable variety; and one specimen, contrasting strongly with the others by its stoutness, might have been from a small modern specimen. All the hoof-bones which I collected, about half a dozen, were very small. Three metacarpals were picked up, all measuring about 7 inches long and 1 inch in breadth at the narrowest part of the shaft; and these are all adult specimens. Two adult metatarsals measure, the one 8 inches in length and the other only 7·3, while in breadth they both measure only ·9 of an inch. A complete adult radius measures only 9 inches in length. A lower end of a humerus is only 2·5 inches broad. Among six calcanea the largest measured 5·5 inches, and the shortest 4·3. In one specimen the orbit is 2·4 inches diameter, and in another 2·8 inches, which is decidedly large.

On the whole, the evidence is to the effect that while the prevalent variety had small horns, and was generally diminutive and slender-limbed, there was mixed with it a variety with larger horns and stouter limbs, whether of greater height or not I cannot say.

Sheep.—Only one portion of horn-core was found with portion of the skull. The portion of horn-core is between 3 and 4 inches long, and at the base its largest diameter is 1·5 inch, its smallest 1 inch. At its inner margin starts at an angle of about 20° from the vertical plane; while I should say that in modern sheep that angle is always 45° at least. I apprehend that this is probably the so-called goat-horned sheep, scarcely now to be got in Shetland.

The following measurements of limb bones may be interesting, as indicating considerable variety in size as well as deviation from modern proportions, as indicated by comparison with the bones of the same sheep skeleton which I have used for comparison in previous communications.

One adult metatarsal measures 5·7 inches long and ·4 broad, and another 5·2 long and ·4 broad at the narrowest part of the shaft. In the modern specimen this bone is 4·8 long and ·5 broad.

Three specimens of adult radius have been gathered, measuring in length respectively 6·6, 6·, and 5·9; while in the modern specimen the corresponding bone is only 5·2.

Two complete humeri are among the specimens gathered. The largest, not quite adult, is 5·7 inches in greatest length; while the other, quite adult, is only 5 inches long, and in the modern specimen the humerus is 5·2 long. Four additional specimens of the lower end of the humerus have been obtained; and one of them is decidedly larger than the largest complete specimen, and another decidedly smaller than the smallest complete bone.

The sheep was therefore long and slender legged, like

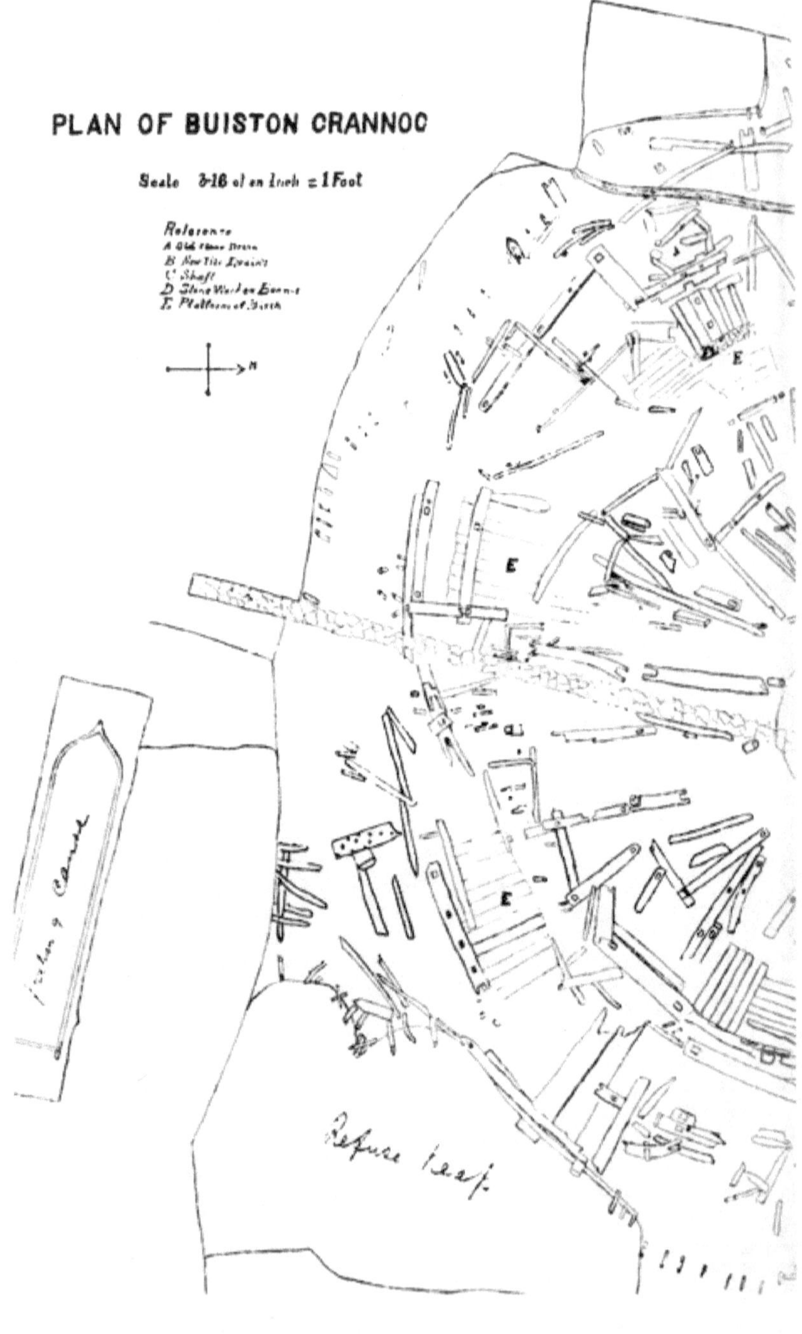

PLAN OF BUISTON CRANNOC
Scale 3-16 of an Inch = 1 Foot
Reference
A Old Stone Drain
B New Tile Drains
C Shaft
D Stone Wall on Beams
E Platform of Beams
N
Refuse heap.

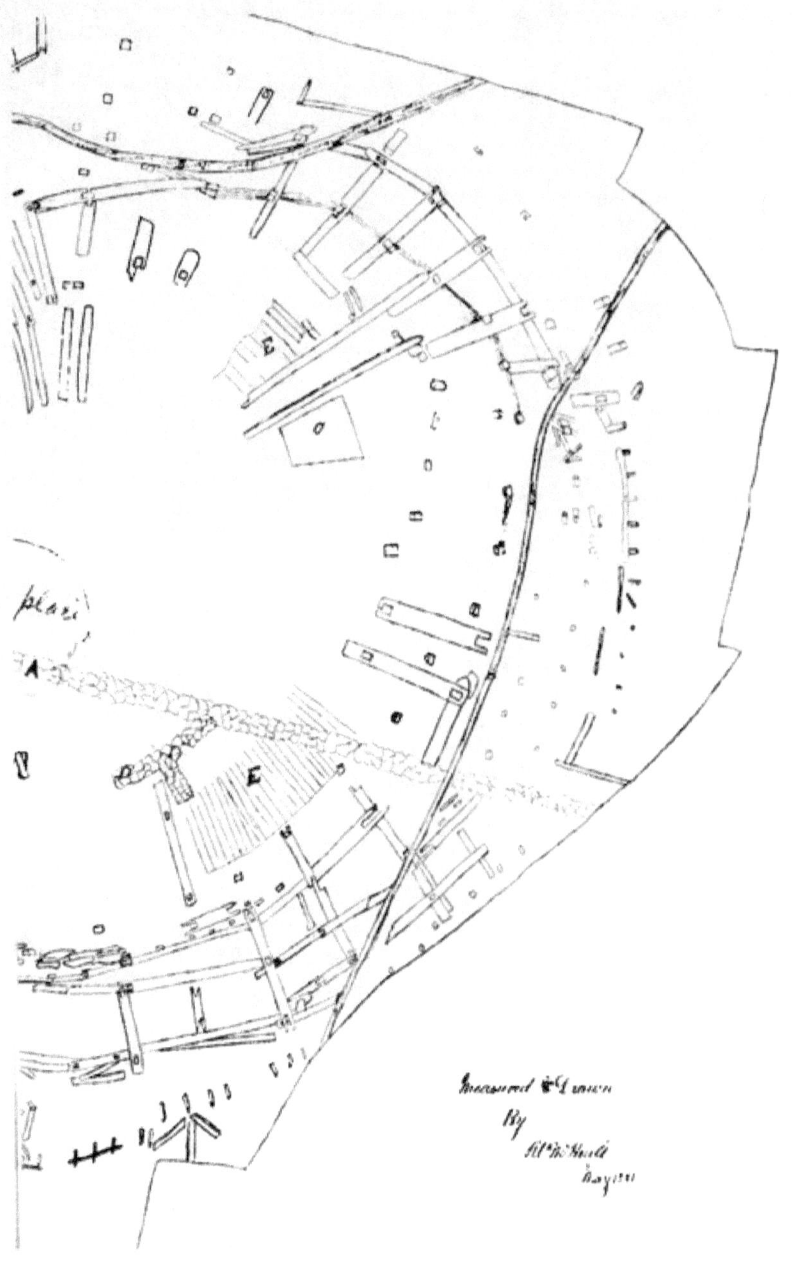
plan
A
E
E
Measured & Drawn
By
Lt. McNeill
May 1811

those found on other Ayrshire deposits.   But it is difficult
to determine whether the differences in size depend on sex,
or some other cause, such as cultivation.

No goat bones have been found in connection with this
lake-dwelling.

CHAPTER V.

GENERAL OBSERVATIONS ON THE CLASSIFICATION, GEOGRAPHI-
CAL DISTRIBUTION, STRUCTURE, AND AGE OF ANCIENT
SCOTTISH LAKE-DWELLINGS.

In the foregoing chapters I have recorded nearly all the facts hitherto derived from the explorations of Scottish Lake-Dwellings, together with a few meagre notices of their former existence supplied by historical research. Notwithstanding the variety and number of objects found in these remains, and the copiousness of details with which the investigations are described, it may still be doubted whether the time has arrived for applying to them the rigid principles of induction, with the view of materially enlarging our knowledge of the early inhabitants of this country. However much variety or novelty may add to the interest attached to antiquarian discoveries, it must never be forgotten that their scientific value is to be determined by the extent to which they can be made to enrich our knowledge of the past phases of human civilisation. While, therefore, fully conscious, on the one hand, of the danger of drawing a series of inferences from too limited an experience, on the other, I feel that to ignore altogether such oft-recurring questions as —When did these lake-dwellings flourish? For what purpose were they constructed? And what grade of civilisation characterised their occupiers?—would be tantalising, if not uncourteous, to general readers who have so far perused the mass of dry details here presented to them. In attempting, therefore, to deal with the scientific aspect of these discoveries, I do not for a moment profess such a minute acquaintance with the science of archæology as to entitle me even to attempt a full exposition of the inferences that may be derived from their careful study and comparison with other antiquarian remains; nor, indeed, do I believe that it is within the province of any one man to give a final decision, as it were *ex cathedra*, on a group or groups of remains that include such comprehensive materials as the products of the art, industry, culture, and social economy of a people existing during an undefined period of time, and lying, in a large measure, outside the pale of our historical records. My purpose therefore is, while endeavouring to gratify the laudable curiosity of general readers, to present archæologists with a rough skeleton, which they are invited, piecemeal fashion, to mould into a shapely figure by their combined and varied experience.

To accomplish this object there are certain historical and other collateral phenomena which, I think, help to circumscribe the general sphere of the problems at issue, and which, therefore, fall to be discussed alongside of the inductions derived from the actual materials now before us. In consequence of the diversity of the phenomena thus appealed to, I have grouped their details under the following sections, by means of which I hope to bring the general effect of their chronological bearing into greater prominence :—

1. Classification and geographical distribution of ancient Scottish lake-dwellings.

2. Historical and traditional phenomena associated with their area of distribution.

3. Mechanical skill displayed in the structure of the wooden islands.

4. Topographical changes in the lake-dwelling area during or subsequent to the period of their development.

5. Chronological, social, and other indications derived from the relics.

Section I.

Classification and Geographical Distribution of Ancient Scottish Lake-Dwellings.

The notices of artificial islands in Chapter II. are confined to such as were found to be constructed on timber or surrounded by stockades. There are, however, many others still extant in several of our Scottish lakes, which appear to be entirely composed of stones and earth irregularly heaped together. In the absence of any historical knowledge as to their age there is no *prima facie* reason why some at least should not be contemporary with the former, as it cannot be assumed that the crannog-builders made wood a *sine qua non* in the structure of islands. There were, no doubt, certain stagnant marshes and small lochs in which a wooden foundation was essential for the formation of an island, owing to the softness and yielding nature of the mud; but, on the other hand, there were others with compact, rocky, or gravelly beds, in which any solid materials, as stones, earth, turf, etc., would be equally suitable. The outlets of the larger lakes, more especially of such as were formed in glacial and rock-cut basins, were more adapted for the latter, and as far as my observations have enabled me to form an opinion, these are the very situations in which the stone islands are now found to prevail. Some of them are mere shapeless cairns, without any indications of having been formerly inhabited, while on others are to be found some remains of stone buildings. As to wooden huts or houses, had such structures been erected over them, it is not likely that they could, for any length of time, have resisted the decaying tendencies of a Scottish climate, so that all traces of them would have disappeared long ago.

The social or military exigencies that led people to construct artificial islands would also lead them to take advantage of such natural ones as would be found most suitable, and we may reasonably infer that it is in the absence of the latter that the former would be resorted to. We have therefore no *prima facie* grounds for dis-associating chronologically the artificially-formed islands of wood or stone, either from each other or from such natural islands as may furnish evidence of early occupancy. The great and primary object of the island-builder was the protection afforded by the surrounding lake or morass, the securing of which has continued to be a ruling principle in the erection of defensive works down to the Middle Ages, long after the wooden islands ceased to be constructed. The transition from an island fort to the massive mediæval castle, with its moat and drawbridge, is but another step in the progressive march of civilisation.

When the greater advantages of stone buildings became generally recognised, the old wooden refuges, so liable to decay, so easily destroyed by fire, and so unsuitable for supporting heavy buildings of masonry, would be gradually superseded. It would then be found easier and better to conduct the water to the stronghold than to construct the stronghold in a natural basin of water, however convenient its locality might be. To the transitional period preceding this great change, which culminated in the almost impregnable moated and mediæval castle, may be assigned many of the remains of stone forts, castles, etc., still abundantly found in bogs, drained marshes, and natural or, maybe, artificially built islands.

While it is therefore possible to assign the wooden islands to a fairly well-defined period, which, speaking generally, precedes that when stone and lime were used for building purposes in this country, the claims of all other island homes to great antiquity must be judged of by their special peculiarities.

The annexed tabular statement comprises not only all the known artificial islands, whether constructed of wood or other materials, but also some natural ones known to have been artificially strengthened or fortified, as well as a few examples of other structural remains, such as camps, castles, etc., now or formerly located in bogs or lakes. The first column of numbers in this table contains only the crannogs proper, *i.e.* islands constructed on wood and surrounded by piles; while the second includes all the remaining classes. The characteristic or differential features of all these examples, when not referred to in the text, will be found in the marginal notes or references, so that a mere glance gives a general idea of their number, character, and geographical distribution.

TABLE showing GEOGRAPHICAL DISTRIBUTION OF LAKE-DWELLINGS, ARTIFICIAL ISLANDS, ETC., IN SCOTLAND, with Notes and References.   Those to which no reference is given will be found described in the text.

COUNTY.	NAME.	Constructed with wool, etc.	Constructed with stones, earth, etc.
Ayrshire,	Loch of Kilbirnie,	1	...
,,	Lochlee,	1	...
,,	Lochspouts,	1	...
,,	Buston,	1	...
,,	Loch Doon,[1]	...	1
(Renfrewshire),	Loch Winnoch (Pail),[2]	...	1
Aberdeenshire,	Loch Canmore,	1	1
,,	Banchory,	1	...
,,	Federatt,[3]	...	1
,,	Peel Bog,[4]	...	1
Buteshire,	Loch Quien,	1	...
,,	Dhu Loch,	1	..
Berwickshire,	Battleknowes,[5]	...	1
Argyllshire,	Kielziebar	1	...
,,	Loch na Mial (island of Mull),	1	...
,,	Ledaig,	1	...
,,	Lochnell,	1	...
,,	Parish of Kilchoman,[6]	...	1
,,	Fasnacloich (Appin),[7]	...	1
Dumfriesshire,	Lochmaben,	1	..
,,	Black Loch of Sanquhar,	1	...
,,	Friars' Carse,	1	...
,,	Loch Orr,[8]	...	1
,,	Lochwood,[9]	...	1
,,	Closeburn,[10]	...	1
,,	Corncockle (Applegarth),[11]	1	...
,,	Morton (parish of),[12]	...	1
Fifeshire,	Collessie,[13]	...	1
,,	Stravithy,[14]	...	1

NOTES AND REFERENCES.

[1] Island, with castle of Saxon and Gothic architecture. Several canoes found near it in the loch.—*New Stat. Account*, vol. v. p. 337.

[2] Old castle on an island, near Castle Semple (*ibid.* vol. xv. p. 69). Canoes also found in the loch.—*New Stat. Account*, Renfrew, p. 97.

[3] Old castle surrounded by a fosse and morass, with access by a stone causeway and a drawbridge.—*Old Stat. Account*, vol. ix. p. 191.

[4] Circular earthen mound, having formerly a wooden castle.—*New Stat. Account*, vol. xii. p. 1089.

[5] Square camp 42 yards each side.—*New Stat. Account*, vol. ii. p. 171.

[6] Small island strongly fortified.—*Old Stat. Account*, vol. xi. p. 281.

[7] Artificial island formed of stones and earth.—*Proc. Soc. Antiq. Scot.* vol. vi. p. 175.

[8] Small island with remains of stone walls.—*Old Stat. Account*, vol. ii. p. 342.

[9] Strong castle in impassable bogs.—*Ibid.* vol. iv. p. 224.

[10] Old castle formerly surrounded by a lake; canoe and bronze tripod found in bed of lake.—*Phil. Trans.* 1756, p. 521; also *Antiq. of Scotland*, Grose, vol. i. p. 150.

[11] Curious wooden structures in moss.—*Proc. Soc. Antiq. Scot.* vol. vi. p. 163.

[12] Old castle, near which canoe was dug up; also a small copper camp kettle and copper teapot.—*New Stat. Account*, vol. iv. p. 96.

[13] Castle in marshy ground.—*Old Stat. Account*, vol. ii. p. 418.

[14] Regular fortalice situated in a bog, with ditch and drawbridge.—*New Stat. Account*, vol. ix. p. 365.

TABLE showing GEOGRAPHICAL DISTRIBUTION OF LAKE-DWELLINGS, ARTIFICIAL ISLANDS, ETC.,
IN SCOTLAND,—*Continued.*

COUNTY.	NAME.	Constructed with wood, etc.	Constructed with stones, earth, etc.	NOTES AND REFERENCES.
Forfarshire, . .	Loch of Forfar, . . .	1	1	[15] *Proc. Soc. Antiq. Scot.* vol. vi. 176.
" . .	Loch of Rescobie,[15] .	...	1	[16] Small island, with traces of stone castle.—*New Stat. Account,* vol. xiv. p. 65.
Inverness-shire, .	Loch Lochy, . .	1	...	
" . .	Loch in Croy (drained), .	1	...	[17] In Loch Moy are two islands, on one of which stands the old resi-
" . .	Loch Gynag,[16] . . .	...	1	dence of the family of Mackintosh. The other is merely a heap of
" . .	Loch Moy,[17] . . .	...	1	stones, probably artificial, and was used by the Lairds of Mackintosh
Kirkcudbrightshire,	Lochrutton, . . .	1	...	as a prison. It is called *Ellan-na-glach.*—*Ibid.* vol. xiv. p. 100.
" . .	Loch Kinder, . . .	1	...	
" . .	Carlingwark, . . .	2	...	[18] Fort surrounded by water. In the drained lake fragments of
" . .	Loch Lotus, . . .	1	...	spears and a silver coin found.—*New Stat. Account,* vol. iv. p. 54.
" . .	Barean, . . .	1	...	
" . .	Borgue (parish of),[18] . .	...	1	[19] Artificial lake, with two islands, said to be seats of Fergus Lord
" . .	Loch Fergus,[19] . .	...	1	of Galloway.—*Old Stat. Account,* vol. xi. p. 25.
Lanarkshire, .	Green Knowe, . . .	1	...	[20] On the north-west border of Loch Spinie there are standing on an
Linlithgowshire, .	Loch Cot, ...	1	...	artificial mound, surrounded by a fosse and drawbridge, the walls of a
Moray, Nairn, and				strong castle called Old Duffus.—*Old Stat. Account,* vol. viii. p. 395.
Elgin, . .	Lochindorb, . . .	1	1	
" . .	Loch Spinie,[20] . .	1	2	[21] Contains an island said to be one of the strongholds of the Wolf
" . .	Loch of the Clans, . .	1	...	of Badenoch; also called Loch-an-Eilean.—*New Stat. Account,* vol. xiii.
" . .	Loch Flemington, . .	1	...	p. 137.
" . .	Loch in Dunty, . .	1	...	[22] Castle situated in a swamp.—*Old Stat. Account,* vol. iv. p. 399.
" . .	Lake of Rothiemurchus,[21] .	...	1	[23] A small island, mostly artificial, with ruins of an old castle.—*Old*
" . .	Mountblairy,[22] . .	...	1	*Stat. Account,* vol. ix. p. 231.
Perthshire, .	Loch Rannoch, . .	1	...	[24] Near each end there is a small artificial island with ruins.—*Ibid.*
" . .	Loch Clunie,[23] . .	...	1	vol. xi. p. 180.
" . .	Loch Earn,[24] . .	...	2	[25] Small island, with ruins of castle.—*Ibid.* vol. x. p. 130.
" . .	Loch Ard,[25] . .	...	1	[26] Middle of loch a cairn of stones.—*Ibid.* vol. xviii. p. 327.
" . .	Loch Laggan, Kippen,[26] .	...	1	[27] *Proc. Soc. Antiq. Scot.* vol. vi. p. 176.
" . .	Loch Morall,[27] . .	..	1	[28] Island, partly artificial.—*Old Stat. Account,* vol. ii. p. 475.
" . .	Loch Tummell,[28] . .	...	1	

County	Loch		
,,	Loch Tay,[29]	...	3
,,	Loch Freuchie,[30]	...	1
,,	Lake in Blairgowrie,[31]	...	1
,,	Moulin (drained),[32]	...	1
,,	Loch Granech,[33]	...	1
,,	Loch Fullah,[34]	...	1
,,	Loch of Monivaird,[35]	...	1
,,	Loch Achray,[36]	...	1
,,	Loch Vennachar,[36]	...	1
,,	Loch Kinnard,[36]	...	1
Stirlingshire,	Loch Lomond,	1	...
Sutherlandshire,	Loch Brora,[27]	...	1
,,	Loch Shin,[36]	...	1
,,	Loch Dolay,[36]	...	1
Ross-shire,	Loch of Kinellan,	1	...
,,	Loch Achilty,	1	...
,,	Loch Glass,[38]	...	1
Roxburghshire,	Castletown,[39]	...	1
,,	Loch of Yetholm,[39]	...	1
Wigtownshire,	Dowalton,	5	...
,,	Loch Inch Crindil,	1	...
,,	Castle Loch,[40]	1	...
,,	Barlockhart Loch,[41]	1	...
,,	Sunonness Loch,[41]	1	...
,,	Barneallzie Loch,[42]	1	...
,,	Machermore Loch,[43]	Several	...
,,	Barhapple Loch,	1	...
,,	Loch Heron,[44]	...	2
,,	Mochrum Loch,	...	1
,,	Fell Loch,[45]	...	1
,,	Merton Loch,	1	...
,,	Eldrig Loch,[46]	3 (?)	...

[29] Several islands.—*Old Stat. Account*, vol. xvii. p. 465, and *Proc. Soc. Antiq. Scot.* pp. 173, 175, and 176; also *New Stat. Account*, vol. x. p. 465.

[30] *Proc. Soc. Antiq. Scot.* vol. vi. p. 173.

[31] In the middle of one of the lakes is a small island, with the remains of an old building.—*Old Stat. Account*, vol. xvii. p. 195.

[32] Castle stood in lake, now drained, with vestiges of a causeway.—*Old Stat. Account*, vol. v. p. 69.

[33] Mr. Robertson's notes.—*Proc. Soc. Antiq. Scot.* vol. vi. p. 177.

[34] *Ibid.* p. 172.

[35] Castle anciently surrounded by water.—*Old Stat. Account*, vol. viii. p. 570.

[36] *Proc. Soc. Antiq. Scot.* vol. vi. pp. 172 to 177.

[37] Small island near the lower end artificially constructed of stones, with ruins.—*Old Stat. Account*, vol. x. p. 363, and *New Stat. Account*, vol. xv. p. 151.

[38] Small island near lower end artificially formed of stones.—*Old Stat. Account*, vol. i. p. 282.

[39] Immense cairn of stones in the midst of an extensive and deep morass. Old castle of Yetholm Loch.—*New Stat. Account*, vol. iii. p. 164.

[40] Contains an island of stones and oak stakes, and mossy bogs on south shore, and a peninsula at north-west, with a double row of stakes.—*Rev. W. Wilson, Glenluce.*

[41] See *Proc. Soc. Antiq. Scot.* vol. x. pp. 737-8.

[42] *Ibid.* vol. ix. p. 397.

[43] *Ibid.* vol. ix. p. 368.

[44] Mr. Faed examined its two islands and found them artificial.—*Ibid.* p. 378.

[45] "On the east shore, opposite Fern Island, I found an oak in the peat, with an axe-mark. My companion waded to the island and reported the remains of a paved ford for 20 or 25 feet next the island."—*Ibid.* p. 378.

[46] "Three crannogs, one with ford to one shore and annular stone heap, the others with a ford to each shore.—*Rev. W. Wilson.*

It is manifest, however, that a table of this kind can have no permanent value, beyond giving a full and accurate statement of discoveries up to date, as further researches may not only change its numerical data, but give a totally different aspect to inferences based on the existence or absence of these remains in certain districts. Thus it is only within the last few years that Ayrshire could be included in the lake-dwelling area, so that, previously, the conflicting statement made by Chalmers,[1] that Galloway was colonised by the Irish about the eighth century, derived some countenance from the archæological discoveries in Loch Dowalton, and other lakes in the neighbourhood, when taken in conjunction with the prevalence of analogous remains in Ireland. Though we cannot, therefore, argue definitely from the present geographical distribution of Scottish Lake-Dwellings, the indications are so clearly suggestive of their having been peculiar to those districts formerly occupied by Celtic races that the significance of this generalisation cannot be overlooked. Thus, adopting Skene's division of the four kingdoms into which Scotland was ultimately divided by the contending nationalities of Picts, Scots, Angles, and Strathclyde Britons, after the final withdrawal of the Romans, we see that of all the crannogs proper, none have been found within the territories of the Angles ;[2] ten and six are respectively within

---

[1] *Caledonia*, Book iii. chap. v. pp. 358 *et seq.*

[2] Amongst the donations to the Museum of the Society of Antiquaries of Scotland, I find various vessels of brass found in marshy ground near Balgone, East Lothian (also deer's horns and bones of animals.—*Proceed.* vol. vi. p. 174), by Sir George Grant Suttie, Bart., which are at least suggestive of the former existence of a lake and a crannog in the locality.

These vessels consist of—

A large bronze (brass) tripod pot, with loops at the neck for handle, 13 inches across the mouth, 15½ inches high, and circumference round the middle 45 inches.

Three other bronze (brass) pots of similar type, varying in size.

A bronze (brass) pot (measuring, etc.).

Shallow bronze (brass) basin, etc.

Portion of a larger bronze (brass) basin.

the confines of the Picts and Scots; while no less than twenty-eight are situated in the Scottish portion of the ancient kingdom of Strathclyde. Nor is this generalisation much affected by an extension of the list, so as to include those stony islets so frequently met with in the Highland lakes. On the other hand, that they have not been found in the south-eastern parts of Scotland may suggest the theory that these districts had been occupied by the Angles before Celtic civilisation—or rather the warlike necessities of the times— gave birth to the island dwellings. In that case we would suppose that their development dates back to the unsettled events which immediately followed the withdrawal of the Roman soldiers, to whose protection the Romano-British population in the south-west of Scotland had been so long accustomed. But this leads me to notice some of the historical phenomena associated with the localities thus referred to.

SECTION II.

Historical and Traditional Phenomena associated with their Area of Distribution.

(Compiled chiefly from Dr. Skene's works.)

In the year A.D. 79, Julius Agricola, with his legions, entered that portion of Britain afterwards known as the

Bronze (brass) tripod vessel, with spout and looped handles, etc.
Bronze (brass) tripod vessel with straight spout, etc.

The following extract of a letter from Sir G. Grant Suttie gives the details of the discovery, dated 16th Feby. 1849 :—"Last autumn my labourers were trenching amongst some rhododendrons in a piece of mossy ground under a peculiar ledge of grey rocks, in my park at Balgone, near my house, and about a mile and a half due south from North Berwick Law, when they found a number of camp-kettles of various sizes, one very large, and in this, one of the goblets was found. They were close to each other, and about 8 feet from the surface. The meadow, extending to about 20 acres, where they were found, was generally under water till imperfectly drained by me ; since then the level has sunk from 3 to 4 feet. I have little doubt that when these kettles were deposited here the meadow was a lake, or at all events a morass."—(*Proc. Soc. Antiq. Scot.* vol. iii. p. 251.)

kingdom of Scotland by way of the Solway Firth, and, quickly subjugating the tribes occupying its northern shore, garrisoned the country as he advanced. The work of the following winter is thus described by Tacitus :—

"To introduce a system of new and wise regulations was the business of the following winter. A fierce and savage people, running wild in the woods, would be ever addicted to a life of warfare. To wean them from these habits, Agricola held forth the baits of pleasure, encouraged the natives, as well by public assistance, as by warm exhortations, to build temples, courts of justice, and commodious dwelling-houses. He bestowed encomiums on such as cheerfully obeyed; the slow and uncomplying were branded with reproach; and thus a spirit of emulation diffused itself, operating like a sense of duty. To establish a plan of education, and give the sons of the leading chiefs a tincture of letters, was part of his policy. By way of encouragement he praised their talents, and already saw them, by the force of their natural genius, rising superior to the attainments of the Gauls. The consequence was, that they, who had always disdained the Roman language, began to cultivate its beauties. The Roman apparel was seen without prejudice, and the toga became a fashionable part of dress. By degrees the charms of vice gained admission to their hearts; baths, and porticos, and elegant banquets grew into vogue; and the new manners, which, in fact, served only to sweeten slavery, were by the unsuspecting Britons called the arts of polished humanity."—(*Vit. Agric.* chap. 21.)

During the following summer, A.D. 80, Agricola pursued his journey northwards and entered on a country hitherto unknown to the Romans, and described by Tacitus as occupied by *new nations*, which he laid waste as far as the Firth of Tay. During the six following years this general was engaged in bringing the wild Caledonians under subjection. In the year 81 he erected a chain of forts between the Firths of Forth and Clyde. Subsequently he visited Argyll and Kintyre, as well as the eastern counties, and explored the interior as far as the Grampian range of mountains. At the same time, the fleet, sailing northwards along the east coast, circumnavigated the island and returned by the straits of Dover to its former station. These exploits roused the warlike spirit of the Caledonians, and all the northern tribes combined to resist the progress of the invaders, the result of which was the famous battle of Mons Grampius (A.D. 86), in which 30,000 Caledonians were totally routed. Tacitus in his description of the battle states that the Caledonians were arranged in lines along the slopes of the rising ground, having their charioteers and cavalry in front, and that their weapons were arrows, long pointless swords, and small targets ; whereas the Romans used large shields and short pointed swords, which gave them the advantage at close quarters. As soon as the battle was known at Rome, Agricola was recalled through the jealousy of the Emperor Domitian, so that no advantage was taken of the campaign, and the result was that the northern tribes beyond the Tay retained their independence.

From the recall of Agricola till the accession of the Emperor Hadrian, A.D. 117, nothing is known of the condition of this part of the island. Hadrian visited Britain in the year 120, and appears to have put down a threatened insurrection by giving up Agricola's line of forts, and limiting the frontier of the province to a line right across the territory of the Brigantes from the Solway Firth to the river Tyne. Along this line he constructed an immense barrier consisting of, first, on the north side, a ditch, then a stone wall, and then an earthen rampart. Between the two latter were roads for the transmission of troops, with stations, castles, and watch-towers. But this barrier did not act for a long time as a check to the independent section of the Brigantes, who, early in the reign of Antoninus, crossed the wall and overran portions of the Roman Province. But Lollius Urbicus, who was sent to Britain, quickly subdued these hostile tribes (A.D. 139), and again extended the Province to its former limit, viz., the line between the Firths of Forth and Clyde, where he constructed an earthen rampart—"Alio muro cespiticio submotis barbaris ducto."

In A.D. 162 an attempt on the Province by the northern tribes was quelled by Calphurnius Agricola.

Twenty years later they made another formidable irruption into the Province, but were repelled by Marcellus Ulpius.

In A.D. 208 the Emperor Severus found it necessary to come in person to repel these frequent and formidable attacks of the northern barbarians, whom we now find under the names of Mæatæ and Caledonii. The former occupied the lower lands next the wall, and the latter the mountainous regions beyond, but notwithstanding the difference in name, they appear to have been virtually the same people.

"The manners of the two nations are described as the same, and they are viewed by the historians in these respects as if they were but one people. They are said to have neither walls nor cities, as the Romans regarded such, and to have neglected the cultivation of the ground. They lived by pasturage, the chase, and the natural fruits of the earth. The great characteristics of the tribes believed to be indigenous were found to exist among them. They fought in chariots, and to their arms of the sword and shield, as described by Tacitus, they had now added a short spear of peculiar construction, having a brazen knob at the end of the shaft, which they shook to terrify their enemies, and likewise a dagger. They are said to have had community of women, and the whole of their progeny were reared as the joint offspring of each small community. And the third great characteristic, the custom of painting the body, attracted particular notice. They are described as puncturing their bodies, so as, by a process of tattooing, to produce the representation of animals, and to have refrained from clothing, in order that what they considered an ornament should not be hidden."—(Skene, *Celtic Scotland*, vol. i. p. 83.)

Severus opened up the country by cutting down woods,

throwing bridges across the rivers, clearing the jungles, and making roads in various directions, and in this manner, after great loss of human life, but without fighting a single battle, he penetrated as far north as the shores of the Moray Firth. Returning through the heart of the High-lands he concluded a peace with the Caledonians, from whom he received hostages. He then rebuilt the wall between the Clyde and the Forth and returned to York. Soon after-wards the Mæatæ and Caledonii again revolted, and thus a second time drew forth the ire of the aged Emperor, but, while he was preparing a severe revenge, death overtook him.

Little is known of the subsequent relative positions of the Romans and Caledonians till A.D. 306, when Constantius Chlorus is said to have penetrated into the low country beyond the wall, and defeated the Picts.

For upwards of fifty years there is again a complete silence as to the conduct of the natives beyond the Roman bound-ary, and it is not till A.D. 360 that they reappear on the historic field. Then the comparative security and prosper-ity enjoyed by the provincial Britons during the last 150 years was broken, and the inroads of the barbarians into the province became so formidable that they appeared to be deliberate attempts to drive the Romans entirely out of Britain. The Picts were now joined by a new nation which emerged from Ireland, and became known to the Romans under the name Scoti. The effect of this combination of hostile tribes is thus described by Mr. Skene :—

"We learn from the account given by the historian of their eventual recovery, that the districts ravaged by the Picts were those extending from the territories of the independent tribes to the wall of Hadrian between the Tyne and the Solway, and that the districts occupied by the Scots were in a different direc-tion. They lay on the western frontier, and consisted of part of the mountain region of Wales on the coast opposite to Ierne, or the island of Ireland, from whence they came. . . . During four years the invading tribes retained possession of the districts they had occupied, and were with difficulty prevented from overrunning the province; but in the fourth year a more formidable irruption took place. To the two nations of the Picts and the Scots were now added two other invading tribes—the Saxons, who had already made themselves known and dreaded by their piratical incursions on the coast; and the Attacotti, who, we shall afterwards find, were a part of the inhabitants of the territory on the north of Hadrian's wall, from which the Romans had been driven out on its seizure by the independent tribes. They now joined the Picts in invading the province from the north, while the attack of the Saxons must have been directed against the south-eastern shore; and thus assailing the provinces on three sides—the Saxons making incursions on the coast between the Wash and Portsmouth, afterwards termed the Saxon Shore, where they appear to have slain Nectarides, the Count of the maritime tract, the Picts and Attacotts on the north placing Fallofandus, the Dux Britanniarum, whose duty it was to guard the northern frontier, in extreme peril, and the Scots penetrating through the mountains of Wales—the invading tribes penetrated so far into the interior, and the extent and character of their ravages so greatly threatened the very existence of the Roman government, that the Emperor (Julian) became roused to the imminence of the danger, and, after various officers had been sent without effect, the most eminent commander of the day, Theodosius the elder, was despatched to the assistance of the Britons. He found the province in the possession of the Picts, the Scots, and the Attacotts, who were ravaging it and plundering the inhabitants in different directions. The Picts, we are told, were then divided into two nations, the 'Dicalidonæ' and the 'Vecturiones,' a division evidently corresponding to the two-fold division of the hostile tribes in the time of Severus, the 'Caledonii' and the 'Mæatæ.' The similarity of name and situation sufficiently identifies the first-mentioned people in each of the twofold divisions. The Mæatæ had been obliged to cede a part of their territory to the Romans, so that part of the nation had passed under their rule, and a part only remaining independent probably gave rise to the new name of 'Vecturiones.' The 'Attacotti,' we are told, were a warlike nation of the Britons, and the epithet applied to the 'Scoti' of ranging here and there shows that their attacks must have been made on different parts of the coast."—(Skene, *Celtic Scotland*, vol. i. pp. 98-100.)

Theodosius, with a powerful army, soon drove back the invaders, and restored the province to its former integrity; but his success was without any permanent result. During the next forty years, till the final withdrawal of the Roman troops in A.D. 410, the provincial Britons, especially those inhabiting the district between the two walls, became a prey to the surrounding hostile nations as often as the increasing demands on the military resources of the Empire at home caused a temporary retirement of its troops. Thus, during the short period here referred to, the portion of the Province was overrun and desolated no less than three different times, and as often restored by the Roman legions. At length, however, a time came when these were destined never to return, and the semi-Romanised Britons were allowed to struggle with the northern barbarians as best they could. What took place in North Britain after this great event, or how the contending nationalities settled the country among them, can only be gleaned by the uncertain voice of tradition; nor was the veil of darkness which thus fell on the country removed till a new source of historical records sprung up by means of the civilising influence of the early Christian Church and her learned emissaries.

"When the page of history once more opens to its annals, we find that the barbaric nations whom we left harassing the Roman province till the Romans abandoned the island, had now effected fixed settlements within the island, and formed permanent kingdoms within its limits. South of the Firths of Forth and Clyde we find her containing a Saxon organisation, and tribes of Teutonic descent, hitherto known by the general name of Saxons, in full possession of her most valuable and fertile districts, and the Romans of the old British provincials confined to the mountains of Wales and Cumbria, the western districts extending from the Solway to the Clyde, and the peninsula of

Cornwall. North of the Firths we find the barbarian tribes of the Picts and Scots, which had so often harassed the Roman province from the north and west, formed into settled kingdoms with definite limits; while Hibernia or Ireland now appears under the additional designation of Scotia."—(Skene, *Celtic Scotland*, vol. i. p. 115.)

The settlements of these four nationalities were as follows:—The Angles occupied the south-eastern district, and ultimately formed the kingdom of Northumbria, which extended from the Firth of Forth to the river Humber. The provincial Britons were divided into several petty states, which originally belonged to two varieties of the British race. Those in the northern districts, corresponding to the Damnonii of Ptolemy, and occupying the modern shires of Lanark, Renfrew, and Ayr, are said to have belonged to the Cornish variety; while the Cymric branch extended as far north as Dumfriesshire. The battle of Ardderyd (Arthuret, west side of the Esk near Carlisle), in A.D. 573, which ended in the defeat of the Angles, consolidated these petty states into the kingdom of Cumbria or Strathclyde under Rhydderch with the fortress of Alclyde (Dumbarton) as its capital. (573-601 Rodercus filius Totail regnavit in Petra Cloithe.—Adamnan.)

The counties of Wigtown and Kirkcudbright were occupied in the second century by the Novantæ, having two towns called Rerigonium and Leucopibia. The ancient Celtic name of the district was in Irish Gallgaidhel (*i.e. forcign Gael*), and in Welsh Galwydel, and hence in Latin *Gallovidia*, *Galloweithia*, now Galloway. Its inhabitants were called by Bede "Niduari Picti," and they were known as the Galloway Picts as late as the twelfth century. The most puzzling statement about this district is that of Chalmers (*Caledonia*, i. p. 358), who states that Galloway

---

[1] See *Proc. Soc. Antiq. Scot.* vol. vi. p. 91.

was colonised in the eighth century by the Cruithne from Ireland. Cruithne is the Irish equivalent to Picti, and a people known by this name occupied the larger portion of Ulster. According to Skene, however, Chalmers's statement is not supported by any evidence.

The Scots forming the kingdom of Dalriada occupied that portion of the west of Scotland corresponding to Argyllshire, and had the fortress of Dunadd as their chief stronghold.

The rest of Scotland, with the exception of a portion of the low country near the Roman wall, which became a debatable territory, and often the theatre of wars amongst the four surrounding nations, constituted the so-called Pictish kingdom.

Christianity was introduced into Scotland from two different sources. The Southern Picts were converted to the Faith by St. Ninian, who derived his teaching direct from Rome, and founded a church at Candida Casa (Whithorn) as early as A.D. 397; while St. Columba, the Apostle of the Northern Picts, came from Ireland in A.D. 563, and developed the Columban branch of the Church, having its headquarters at Iona. The more important of the subsequent events of these four kingdoms are here briefly arranged in chronological order :—

A.D. 573. Battle of Ardderyd.

A.D. 575. Aidan becomes king of the Scots.

A.D. 603. Angles of Bernicia defeat a combined army of Britons and Scots under the command of Aidan at Degsastane, now Dawstone, in Liddesdale.

A.D. 606. Death of Aidan, king of Dalriada; Aedilfrid conquers Deira and expels Acduin.

A.D. 617. Acduin regains the kingdom of Northumbria, and eleven years afterwards he and his people are converted to Christianity by Paulinus.

A.D. 642-670. Angles, under King Oswy, subdue and make tributary to him the Britons of Strathclyde, as well as the greater portion of the Picts and Scots.

A.D. 672. Unsuccessful attempt of the Picts to throw off the yoke of the Angles.

A.D. 684. Ecgfrid, king of Northumbria, sends an army to Ireland, and lays waste part of that country. In the following year he invades the kingdom of the Picts, but is defeated and slain at Dunnichen. The Picts, Scots, and Britons of Strathclyde, etc., now regain their freedom, but the Angles still retain possession of Galloway.

A.D. 740. Alpin, king of the Scots of Dalriada, invades Galloway, but is slain near Kirkcudbright.

A.D. 744. Battle between Angus, king of the Picts, and the Britons of Strathclyde. Soon afterwards the former is joined by Eadberet of Northumbria, and a combined attack on the kingdom of Strathclyde is made, with the result that the latter adds the whole of Ayrshire to his Galloway possessions. (Eadbertus campum Cyil cum aliis regionibus suo regno addidit.—Bede, *Chron.*)

A.D. 795. First appearance of Norwegian and Danish pirates in the western seas.

A.D. 802. Iona burned by Norsemen.

A.D. 806. Iona again plundered by Norsemen, and sixty-eight men of the monastery slain.

A.D. 844. Kenneth mac Alpin, king of the Scots, becomes also king of the Picts.

A.D. 853. Arrival of Olaf the White in Ireland. He seizes Dublin, and establishes himself there as king, after which he makes an expedition into Scotland, besieges, and takes Alclyde after a siege of four months.

A.D. 946. Kingdom of Cumbria ceded to Malcolm, king of the Scots. (Strat Clut vastata est a Saxonibus.—*Hist. Brit.*)

SECTION III.

Structure of the Wooden Islands.

In my Introductory Chapter I have remarked that none of the Irish writers appear to have paid much attention to the mechanical principles on which the wooden islands were constructed. A similar remark is equally applicable to the writers on Scottish crannogs. Dr. Stuart had got hold of the general idea that the mortised transverses were used for the purpose of steadying the uprights, and that the outer structures were adapted to resist the action of the surrounding water. The following are his words: "Of the first class, or the crannog proper, the ordinary construction was by logs of wood in the bed of the lake supporting a structure of earth or stones, or of a mixture of both, the mass being surrounded by piles of young oak-trees in the bed of the lake, the inner row of which kept the island in shape, and the external rows acted as defences and breakwaters."[1] But these views convey only a partial notion of a more comprehensive system, the meaning of which I was only able to perceive after my experience at Buston. Notwithstanding that I made the structural arrangements of the Lochlee crannog a particular point of study, I failed to adduce a satisfactory theory for the details recorded; and, beyond showing that the two inner circles of uprights, with their radial and circumferential transverses which immediately surrounded the log pavement, formed a kind of breastwork or wall some 3 feet high, I made no advance on Dr. Stuart's theory. I could offer no explanation of the other large mortised beams found external to this circle, nor of the network of oak

---

[1] *Lake-Dwellings*, by Keller, Second Edition, p. 657.

beams—some with mortised holes, and others with tenons to fit into continuous beams—which became manifest on making the deep shaft, and appeared to permeate the whole structure of the island. The advantage of carefully recording every fact, however trivial or obscure, has never been better illustrated than in this very article (that on the Lochlee crannog), as, with the more recent light thrown upon the subject, there can be no doubt that these structures, as well as the mortised beams (one of which contained three holes) lying on the western margin of the crannog, were some of the radial beams of an encircling girdle, still *in situ*, which surrounded and knit together the island in a precisely similar manner to that at Buston, as described at page 197. At Buston also, the inner circle, as evidenced by its mural remains, formed part of the enclosure surrounding the log pavement, and thus corresponded with, and served the same purpose as, the breastwork at Lochlee. Again, on the south side of both crannogs, the circles were more numerous, and occupied a larger area than on the north side, but with this difference,—that at Lochlee the midden covered the greater portion of this space, which at Buston was converted into an open and partially paved promenade.

In the incidental notices of these islands, given in Chapter II., the narrators sometimes describe them as having been built on a framework of oak, as at Lochrutton, Loch Kinder, etc. At other times the only evidence of a crannog consists in fishing up oak beams from the bottom of a lake, or their discovery under its surface inextricably mortised into others, as at Loch Lomond, Lochmaben, Loch Lochy, etc., while in drained lakes and morasses the chief indications are the tops of upright piles surrounding a flat mound. But all these accounts, as well as the more recent notices of crannogs, are characterised by two prominent structural features, viz., (1) upright piles in the form of one or more circles; and (2) the remains of flat beams containing large square-cut holes at their extremities.

The quotations from the Irish writers, given at pp. 6-11, will also show that the same structural features characterise many of the descriptive notices of the crannogs of Ireland. Dr. Stuart also draws particular attention to similar beams at Dowalton, some of which were of "great size and length (one of them 12 feet long), with three mortised holes in the length, 7 inches square." These indications, as well as a careful study of the structural details recorded in my reports of the explorations made at Lochlee, Lochspouts, and Buston, have led me to believe that all the wooden islands were constructed after one uniform plan, and that this plan was actually the outcome of the highest mechanical principles that the circumstances could admit of. Indeed, I am prepared to maintain that were the same problem submitted to modern engineers, they could make no improvement either on the principle or mechanism displayed in these singular structures. Let me therefore, in the first place, note the conditions that called forth such high mechanical and engineering qualities. For defence and protection, which, I presume, no one will doubt were the primary objects of these islands, a small mossy lake, with its margin overgrown with reeds and grasses, and situated in a secluded locality amidst the thick meshes of the primeval forests of those days, would present the most desirable topographical conditions. Having fixed on such a locality, the next consideration would be the selection of materials for building the island. In a lake containing the soft and yielding sediment due to decomposed vegetable matter, it is manifest that any heavy substances, as stones and earth, would be totally inadmissible owing to their weight, so that solid logs of wood, provided there was an abundant supply at hand, would be the best and cheapest material that could be used. To construct in 10 or 12 feet

of water, virtually floating over an unfathomable quagmire,
a solid compact island, with a circular area of 100 feet or
more, and capable of enduring for centuries as a retreat for
men and animals, would, I daresay, be the means of eliciting
from many an engineer of the present day a more frequent
manifestation of the proverbial symptom of a puzzled Scotch-
man than from these early brothers of the craft—the crannog-
builders.

This is how they worked :—

(1.) Immediately over the chosen site a circular raft of
    trunks of trees, laid above branches and brush-
    wood, was formed, and above it additional layers
    of logs, together with stones, gravel, etc., were
    heaped up till the whole mass grounded.

(2.) As this process went on, upright piles, made of oak,
    and of the required length, were inserted into
    prepared holes in the structure, and probably also
    a few were inserted into the bed of the lake.

(3.) The rough logs forming the horizontal layers were
    made of various kinds of wood, generally birch, it
    being the most abundant. These were occasion-
    ally pinned together by thick oak pegs, and here
    and there at various levels oak-beams mortised
    into one another stretched across the substance of
    the island, and joined the surrounding piles.

(4.) When a sufficient height above the water-line was
    attained, a prepared pavement of oak-beams was
    constructed, and mortised beams were laid over
    the tops of the encircling piles which bound them
    firmly together as already described. The margin
    of the island was also slantingly shaped by an
    intricate arrangement of beams and stones, con-
    stituting in some cases, according to Dr. Stuart,
    a well-formed breakwater.

(5.) When the skeleton of the island was thus finished, probably turf would be laid over its margin where the pointed piles protruded, and a superficial barrier of hurdles, or some such fence, erected close to the edge of the water.

(6.) Frequently a wooden gangway, probably submerged, stretched to the shore, by means of which secret access to the crannog could be obtained without the use of a canoe. (For the mode of structure of gangway, see page 101.)

Bearing in mind that all these structures were solidly put together without nails or bolts, and that the gangways, which have remained permanently fixed to the present time, had neither joint nor mortise, we may fearlessly challenge modern science to produce better results under these, or indeed any, circumstances.

That future discoveries may show slight deviations from the exactness of these details I am quite prepared to believe, as the mechanical knowledge of the age was too thorough not to be readily adapted to varied circumstances. What I do however maintain is, that the general principles of structure here laid down, being the outcome of a sound knowledge of mechanics, varied during the whole period of their practical application as little as the essential laws of architecture do in the structure of different styles of our modern houses.

Not only do these wooden islands evince much mechanical and technical skill on the part of their founders, but, what is still more singular, their area of distribution appears to have been co-extensive with the districts formerly occupied by Celtic races. Hence we have here another proof of the extraordinary vigour, intense individuality, and plastic character of early Celtic civilisation, in thus developing, from its own inherent resources, a unique form of stronghold, simple

in structure, but yet admirably adapted to the unsettled conditions of life and military requirements of the period.

SECTION IV.

*Topographical Changes in the Lake-Dwelling Area during
or subsequent to the Period of their Development.*

Supposed Change in Climate, and its effects.—The argu-
ments in support of the theory that a material deterioration
has taken place in the climate of Britain since the Roman
period are generally based on the following considera-
tions :—

(1.) *Historical Statements.*

Roman historians agree in representing the climate as humid,
but so mild that the natives went about in a semi-nude condition,
favourable to luxuriant vegetation, and not so cold as that of
Gaul. Thus Cæsar expressly states that all kinds of trees grew
in Britain, except the *fir* and the *beech*, and that the climate was
more temperate than in Gaul. Tacitus describes the climate
as always damp with rains, and overcast with clouds, without, however, intense cold being even felt. In the speech attributed to Galgacus, previous to the battle of Mons Grampius, he
says that their " bodies and limbs were worn out by cutting
down woods, draining bogs, stripes, and outrages ; " and the
same historian, in describing the result of a previous engagement
in Fifeshire, in which the Caledonians were beaten, states that
" had not the woods and marshes favoured their retreat, this
victory in all probability would have put an end to the war."
Another writer (Dion Cassius) describes the Caledonians as dwelling in tents, naked, and without shoes ; enduring hunger, cold,
and all manner of hardships with wonderful patience ; and
capable of remaining in bogs for many days immersed up to the
neck, and without food. In the woods they lived on the bark of
trees and roots, and had a certain sort of food always ready, of
which, if they took but the quantity of a bean, they would
neither be hungry nor dry for a long time after. Herodian also
describes them as going about partially naked to prevent the
beautiful figures painted on their bodies from being hidden.

According to him, they used iron as other barbarians did gold, both as an ornament and sign of wealth. They wore neither coat of mail nor helmet to prevent being encumbered in their marches through bogs and mosses, whence such a quantity of vapours was exhaled that the air was always thick and cloudy.

(2.) *The numerous remains of forest trees found in bogs and mosses in localities where, at the present time, no such trees are found to grow.*

As indicating the kind of evidence brought forward in support of this argument, the following extracts will suffice:—

"Of old, in the parish of Croy, Inverness-shire, and before the records of the kingdom, there were extensive forests of oak, birch, fir, and hazel, which have been converted into moss, in some places upwards of 20 feet deep. In a moss 400 feet above the level of the sea, oaks of extraordinary size are dug up, some of them measuring from 50 to 60 feet, and of proportional thickness; and even at the height of 800 feet, where the parish joins the Strathdearn hills, large blocks of fir are found, where now, from cold and storm, the dwarf willow can scarcely raise its downy and lowly head."—(*New Stat. Account*, vol. xiv. p. 449.)

A writer in the *Old Stat. Account* (vol. xv. p. 484) referring to moss in the parish of Kilbarchan, Renfrewshire, from 7 to 9 feet deep, says:—

"The soil below is a deep white clay, where has formerly been a forest. The oak is perfectly fresh; the other kinds of timber are rotten. The stumps in general are standing in their original position. The trees are all broken over at about the height of 3 feet, and are lying from south-west to north-east. So, wherever you see a stump, you are sure to find a tree to the north-east. How an oak-tree could break over at that particular place, I never could understand. But we may be allowed to form a conjecture, that before the tree fell, the moss had advanced along its stem and rotted it there."

Mr. Aiton, in an excellent introduction to his treatise on Moss Earth, thus writes:[1]—

---

[1] *A Treatise on the Origin, Qualities, and Cultivation of Moss Earth*, 1811.

"Trees of enormous dimensions have grown spontaneously in many parts of Britain, where it would baffle the ingenuity of man to rear a tree to the tenth part of the size. The mosses in all parts of the island abound with trees of much greater dimensions than any now to be found growing in this country. The late Mr. Browning found an oak-tree under a moss in his lands of Benthall, in East Kilbride, of such size and preservation as to floor a garret 20 feet long by 16 wide. The boards, more than an inch in thickness, may still be seen at Benthall. Another oak-tree may be seen there upwards of 60 feet in length. It has evidently been broken at both ends, and the lower end not being completely covered with moss, has rotted so much that the dimensions of the tree cannot now be ascertained, but the upper end is more than 4 feet in circumference.

" At Thriepwood, in Dalserf parish, and county of Lanark, the trunk of an oak-tree 65 feet in length was a few years ago dug from under moss. It was as straight as the mast of a ship, and so equal in thickness at both ends, that it was not easy to say which was the root. Both these trees had grown at 500 feet of altitude above the level of the sea, and on ground on which it would be difficult to rear an oak to the twentieth part of the size.

" Many fir-trees 100 feet in length have been found under moss. But, what is still more surprising, oak-trees 100 feet long were found on draining Hartfield Moss in Yorkshire. They were as black as ebony, and some of them sold one hundred and fifty years ago as high as £15 for one tree. One oak-tree was dug from under that moss, which measured 120 feet in length, 12 feet in diameter at the root, and 6 feet diameter at the top ! Twenty pounds sterling was then offered for that single tree. One of the same dimensions, with its bark, would sell now at £300, but no such tree at present exists in Europe. . . . .

" But the climate was then also more congenial to the growth of fruits, grapes, etc., as we have seen, and also of grain, than it is now. The records of the Religious Houses show that wheat was paid as a tithe from lands on which human industry could not now raise that species of grain. Wheat was paid annually as a tithe to the Priory of Lesmahagow from lands in that parish, on which that species of grain has not been sown for several centuries past, and where it could not now be raised ; where, under the present economy, oats can scarcely be brought to perfection.

The minister of Glenluce, in the Statistical Account of his parish, mentions a farm that paid to the monastery of Glenluce twelve bolls of wheat and the same quantity of barley as a tithe. But such is the melancholy alteration that, about thirty years ago, that very farm was set at £12 of yearly rent. Similar instances might be pointed out in many parts of Scotland."

The inferences derived from these and similar observations made on the phenomena of peat bogs and their buried forests in many other parts of Scotland, are somewhat conflicting. There can be no doubt that the climatal conditions that permitted oaks to grow on the higher slopes of the hills in the north of Scotland, as well as trees of considerable size in the Orkney and Shetland Islands, where scarcely a stunted shrub is now to be seen growing in a wild state, were more favourable to the growth of forest trees than those which now prevail. On the other hand, the large pines found in some of the Lowland mosses would seem to indicate a colder climate. The probable explanation of this is, that the pines and oaks belong to different periods of time, that the former preceded the latter and flourished when Scotland stood higher above the level of the sea, and, being thus more extensive, was under a colder climate, somewhat analogous to that which now prevails in Norway. These conditions gradually gave place to a more genial climate, which also induced a corresponding change in its flora. This apparently more temperate climate appears to have been succeeded by another change in the climatal conditions, this time of a less genial character, and not so favourable to the growth of the oaks, and to which they, in their turn, gradually yielded in point of luxuriance, and ultimately succumbed to the encroachment of the bogs and peat-mosses.

To discuss at greater length the climatal changes which may have occurred since the great oaks and pines of our

moss-covered forests flourished, would be a digression from
the main object of this work. In geological and astrono-
mical causes some scientists believe they have a sufficient
explanation of the strange, and sometimes severe, oscilla-
tions that are known to have taken place in post-Tertiary
times, so that the disappearance of the ancient forests
from the country, and the subsequent increase of the peat
bogs, may be looked upon as the effect rather than the
cause of our altered climate. There can hardly be any
doubt, however, that during prehistoric times, and towards
the dawn of European history, North Britain was extensively
covered with forests which were characterised by a luxuriance of growth which is not now attained in the same
latitudes. Hence the statements made by Roman historians
to the effect that the natives painted their bodies and
roamed about in a semi-nude condition must not necessarily be stigmatised as fables. That the inhabitants themselves greatly contributed to the clearance of the woods
and jungles, as they became practically acquainted with
the advantages of systematic tillage of the ground, is
also probable. But this change was not produced all at
once, as it is proved by such numerous place-names as
Woodlands, Woodend, Woodside, Linwood, Fulwood, Oak-
shaw-side, Oakshaw-head, Walkingshaw, Lainshaw, etc.
etc., that the south-west of Scotland was well wooded
after the Celtic language had been superseded by the
Saxon dialects. Cosmo Innes thus alludes to the subject: "At the earliest period illustrated by the Melrose
charters there is sufficient evidence that the southern division
of Scotland was not a well-wooded country. On the contrary,
the right of cutting wood was carefully reserved when
pasturage or arable land was granted; and when that right
was conceded for some particular purpose, such as for fuel
for a salt-work, or for building, the use was limited in

express terms. The high grounds of Ayrshire may be an exception, where there seems to have existed an extensive forest; but elsewhere wood was a scarce and valuable commodity."[1]

Though the gradual extinction of our primeval forests, the growth of peat bogs, and many other topographical phenomena, are thus distinctly traceable in the long vista which extends backwards to prehistoric, if not, indeed, to geological times, there are no definite landmarks, beyond a certain chronological sequence, by means of which these physical changes can be more directly measured on the scale of time.

Amidst the abundant traces of struggling humanity, by which the whole line of this hazy vista is characterised, we see the remains of these lake-dwellings. The topographical features and environments of Loch Buston, when the crannogbuilders commenced their structural operations, were totally different from what they are now. Then a stagnant lake, deeply encroached upon by a marginal zone of aquatic plants, and surrounded by a forest of oaks[2] and other indigenous trees, occupied the site of the present fertile basin; now the whole country-side is laid out into regular and well-cultivated fields, with encircling hedges, and scarcely a tree to mark the once wooded locality, part of which is still significantly known as the Shaws. For upwards of half a century the present farmer has been raising splendid crops of grain, not only from the whole area of the dried-up lake, but over the very site of the crannog. And as for the size of the trees which formerly grew here, I question if there is an oak now growing in the whole of Ayrshire from which a dug-out canoe, having dimensions as large as the one found at Loch Buston, could be made. The scene strongly reminds me of

---

[1] *Sketches of Early Scottish History*, p. 100.

[2] Portion of the trunk of an oak-tree, recently dug out of the moss at the margin of the lake basin, still lies on the surrounding hedge bank.

a visit to Ephesus, when, after inspecting the ruins of its once busy harbour, I penned the following words :—

" This (the famous Panormus) consists of a stone-built quay, overlooking what used to be the harbour, having behind it a series of vaults and passages, which must have been used as stores. Now, however, instead of gazing on the sea, studded with ships from all nations, and a crowded harbour, as St. Paul did when he landed at Ephesus, we had before us a rich green field of wheat just coming into ear, dotted here and there with some ugly-looking fig-trees. As for the sea, it was nowhere to be seen, being distant some four or five miles. The alluvial deposits of fifteen centuries have thus not only raised the general level of the valley, and covered it with débris to the thickness of about 12 feet, but also very considerably enlarged its area, and converted it now into an unhealthy and marshy wilderness." [1]

Increase of Lake Silt.—But is there nothing in the local phenomena of these lake-dwellings to indicate, even approximately, the period of their existence, or the changes that have taken place since, by submergence, they have disappeared from the gaze of mankind ? Dame Nature retains many agents in her service who faithfully keep tally of many passing events, though not always by days or years. The woody rings of a tree, water-worn channels, strata in rocks, and accumulated mud, are some of the piles of records which she freely places at our disposal—though often only to baffle our limited and feeble efforts to decipher them. Can we therefore elicit any reliable data from an examination of the accumulated silt from which the remains of the lake-dwellings are now and again disentombed ? Let us see. A few yards from the Buston crannog, where the canoe was found, the upper stratum of the lake sediment, for about the thickness of 2 feet, consisted of a brown clayey substance, but, underneath this, the stuff had assumed a more peaty character, being softer, darker in colour, and

---

[1] *Notes of a Tour in the East*, 1875.

not so heavy. The line of demarcation between these layers was so well marked, that Mrs. Anstruther, in taking a sketch of the canoe while still lying in its original bed, shows this feature on the exposed wall of the trench (Fig. 191). The lowest portion of the canoe was exactly 6 feet below the surface, so that, since its final abandonment, 4 feet of this dark substance gradually accumulated around it. Then however, some sudden change appears to have taken place in its composition, and under the altered circumstances other two feet were added.

It is strange that this phenomenon is identical with what has been observed and recorded at Lochlee. In making a large trench, so as to expose the deeper structures of the gangway, the following particulars were noted (see page 100) :—

(1.) The uppermost portion was a bed of fine clay rather more than 2 feet thick.

(2.) Below this was a soft dark substance, formed of decomposed vegetable matters.

(3.) At a depth of 7 feet the first horizontal beams of the gangway were encountered.

(4.) At 10 feet deep the base of the gangway was supposed to be reached.

The canoe found at Lochlee, though in a different part of the lake-bed, was also buried to a depth of from 5 to 6 feet.

Regarding the marked change in the upper portion of the sedimentary deposits in both these lakes, perhaps no better explanation can be offered than that already suggested at page 151, viz., that it is due to the disappearance of the primeval forests which formerly covered their drainage areas, and the conversion of the land to agricultural purposes, when part of the broken-up soil and clay would be washed down into the lake-basin, and so become mixed with ordinary silt. Another possible explanation may be found in the gradual filling up of the lakes, and their entire conversion into marshes, covered with rank grasses and other aquatic plants whose decomposition would henceforth take the place of their former mossy sediment.

The proximity of Buston and Lochlee, as well as the topographical similarity in the surrounding landscapes, would indicate that this great change took place about the same time in both these localities. Long before this, however, both canoes were finally sunk or abandoned, the interval of time being measured by no less than 4 feet of a deposit. Again the entire time that has elapsed since the gangway at Lochlee was laid down till the lake was drained, and tally ceased to be kept, is represented by 10 feet.

Subsidence of the Crannogs.—Before quitting this somewhat speculative line of research, one other subject remains to be discussed, viz., the submergence of the Lake-Dwellings. This phenomenon is a uniform feature in all those hitherto examined, and though the causes of it have not been much inquired into by previous writers, they will, I think, be found of sufficient importance to merit the attention of both the antiquary and the geologist. There are just two immediate causes to which this result can be assigned : viz., either a subsidence of the surface of the island, or a rising of the waters of the surrounding lake ; or, as may happen sometimes, a combination of both causes. The physical agencies that are likely to operate in producing a subsidence of the island may be categorically stated thus :—

(1.) A compression of the island due to consolidation or decay of its structural materials.

(2.) A sinking of the whole mass of these materials in soft mud as a direct result of weight.

(3.) A general compressing and sinking of the sedimentary strata of the lake-bed.

When the deeper structures of the Lochlee crannog were

examined, it was particularly noted that there was no flattening of any of the large logs, as if they had been subjected to
great pressure. At Lochspouts and Buston, so far as the
water permitted of similar explorations being made, this
observation was equally applicable to them. Though quite
soft, the logs always preserved their original shape and contour. One day, at Lochspouts, I was greatly puzzled by
finding what was evidently portion of a birch-tree, from
6 to 9 inches in diameter, quite flat, and with scarcely any
wood left inside the thick bark. In no instance previously
had I seen the evidence of pressure on logs of this size; but
after carefully considering the point, it was ascertained that
such effects occurred only in the upper portion of the mound,
and above the log pavement, where the wood had been exposed to atmospheric influences, so that when the woody
fibres rotted away, the flattening of the bark was easily produced. All the logs found buried in water or mud retained
their original dimensions, and showed no traces of having
yielded to superincumbent pressure.

In calculating the pressure of the entire crannog on the
lake bottom, it is only necessary to take into account the
weight of the materials above the surface of the island, as the
greater density of the displaced water would act as an up-
ward pressure sufficient, before the mass attained its equili-
brium, to allow the surface of the island to project a few feet
above the level of the water, the amount varying according
to the depth of the lake. After the island grounded, if constructed on the principles suggested at page 262, any additional structures would act as a direct weight on the bed of
the lake; but in estimating the final and total submergence
due to this element, we must consider the weight of rubbish
gradually accumulated during the period of occupancy. As
the base of the island at Lochlee was only 4 feet below the
level of that of the gangway, it follows therefore that the

maximum result due to the weight of the island could not exceed this amount.

But the most important cause of submergence is the gradual compression or consolidation of the lake-bed due to the increase of its sediment. The depth of this increase at Lochlee, at least since the gangway was laid, was found to be not less than 10 feet. Independent altogether of any chemical changes going on in the sediment, however gradually formed, its own weight must have acted very considerably in pressing the lower strata into less bulk. Its accumulated depth, however, is far from giving a correct indication of the rise in the bed of the lake; in fact, 10 feet of silt might not raise the latter to half this extent. Another thing, which must not be forgotten in this discussion, is the subsidence which takes place after bogs and marshes are drained. This is a fact well known to those conversant with the effects of drainage. In the *Old Statistical Account of Scotland*, vol. xi. p. 163, I find it stated that in three years after the drainage of Kinordy Moss, its surface sank 3 feet; and Sir George Grant Suttie, writing of a drained marsh at Balgone, says that after drainage its level sank from 3 to 4 feet (see footnote, page 249). Bogs are in fact like sponges saturated with water, swollen to such an extent that they occupy a much larger space than their solid materials would otherwise do. It will also be remembered (page 191) that, after the last and more careful drainage of Loch Buston, some five years ago, the subsidence of the *Knowe* was sufficiently noticeable to attract the attention of the farmer.

To assign more accurately to these agencies the respective amount of subsidence due to each is impossible, but that their combined effect is sufficient to account for the total submergence of the principal lake-dwellings hitherto examined is proved by the measurements and observations made at Lochspouts (page 168), which show a minimum result of 10 feet.

If the above reasoning be correct, little importance remains to be attached to the rising of the waters as a cause of submergence, even in the exceptional circumstances where the agencies that produce this effect, such as the destruction of the forests and the increase of peat bogs, are known to have been in operation. It occurs in localities where the outlet is level and the flow of water sluggish. Those of the Irish writers who have taken notice of the phenomenon generally assign it to this cause. Sir W. R. Wilde thus refers to it: "We likewise learn from their recent submerged condition how much water had accumulated on the face of the country since their construction, probably owing to the great decrease of forest timber and the increased growth of bog. From the additions made to the height of the stockades, and also from the traces of fire discovered at different elevations in the sections made of these islands, it may be inferred that the rise of the waters commenced during the period of their occupation."—(Wilde's *Catalogue*, p. 221.)

The observations made at Lochlee led me to ascribe an exceptionally large share to this element as a cause of the submergence of the crannog, but since then further investigations have proved that the phenomenon takes place to a similar, if not greater, extent in localities where no rising of the surface of the waters could have occurred.

SECTION V.

*Chronological, Social, and other Indications derived
from the Relics.*

Having thus glanced over the collateral phenomena bearing on the age or period of the Scottish Lake-Dwellings, it only remains to say a few words on the general evidence supplied by the relics themselves. We have already seen

that the scattered remains of the artificial islands brought to
light in modern times extended over a large area, embracing
nearly the whole of Scotland, with the exception of its two
northernmost counties and a few others lying on its south-
eastern extremity.   So far, however, as the discovery of
actual remains illustrative of the civilisation and social condition of their occupiers is concerned, we are almost entirely
limited to the results of the investigations made at Dowalton,
Lochlee, Lochspouts, and Buston, all of which are within the
counties of Ayr and Wigtown.   In instituting a comparison
between the contents of these four groups, their analogy, not
only as regards the structure and local distribution of the
islands, but as regards the general character of the relics, is
so wonderfully alike that we have no difficulty in assuming
that originally these lake-dwellings within this area were
erected by one and the same people for a special purpose,
and about the same time, or at least within a comparatively
limited period.   It is true that a considerable diversity exists
as regards the number and character of the relics found in
some of these localities ; thus neither Samian ware nor
implements of bone or horn have been found at Dowalton,
though these are amongst the relics from all the Ayrshire
examples. Such negative evidence, however, does not amount
to much, more especially in this case, as the absence of
Samian ware is more than compensated for by the presence
of other articles presumably of Roman origin; and, moreover, I believe that a more careful search would have greatly
increased the number of relics from Dowalton.

This uniform similarity in these remains, though not
entitling us to extend the above generalisation beyond a certain geographical area in the south-west of Scotland, is however sufficiently marked to enable us to dispense with any
further necessity of discussing the merits of each group
separately ; so that whatever inferences can be legitimately

derived from a critical examination of any one group may be safely applied to the whole.

If these observations are really trustworthy, we may at once proceed to examine the remarkable series of implements, weapons, ornaments, and nondescript objects here presented to us, with the view of abstracting from them some scraps of information regarding their original owners. Fragments of Samian ware (Fig. 180), bronze dishes (one with Roman letters, Fig. 13), harp-shaped fibulæ of peculiar type (Figs. 140 to 143), together with a large assortment of beads, bronze and bone pins, bone combs, jet ornaments, etc. etc., are so similar to the class of remains found on the excavated sites of Romano-British towns that there can hardly be any doubt that Roman civilisation had come in contact with the lake-dwellers and partially moulded their habits. The Celtic element is however strongly developed, not only in the general character of many of the industrial implements of stone, bone, and iron, but also in the style of art manifested in some of the ornamental objects included in the collection. Thus the piece of wood (Figs. 149, 150) with its carved spiral patterns, the combs, especially the one represented by Fig. 30, showing a series of concentric circles connected by a running scroll design, the table-man carved with similar circles and an open interlaced knot-work; and the bronze brooch (Fig. 26), present a style of ornamentation which is considered peculiar to early Celtic art. The spiral finger-rings seem also to have been of native origin, and the probability is that they were manufactured where they were found, as several crucibles are amongst the relics from the same lake-dwelling, one of which, from the fact that it still contains particles of gold, unmistakeably demonstrates that it had been used in melting this metal.

On the other hand, the forged gold coin (Fig. 216) is the only relic that can with certainty be said to have emanated

from a Saxon source; at least, that cannot otherwise be
accounted for.

But if, from internal evidence, a presumptive case is
made out in favour of the Celtic origin and occupation of
these lake-dwellings, it is greatly strengthened when we

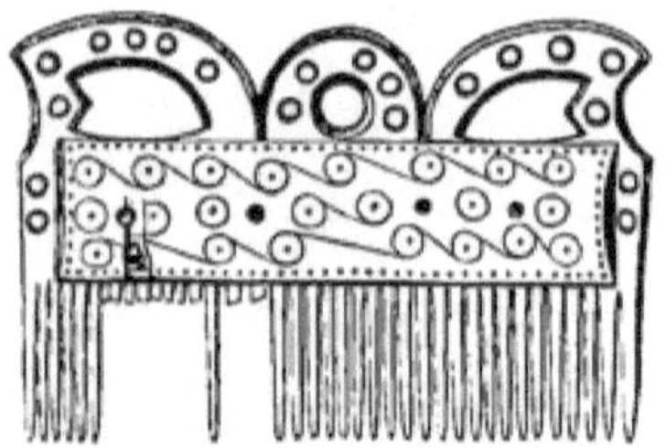

Fig. 255.—Bone Comb, from Ballinderry
Crannog, Ireland (2½ inches long).

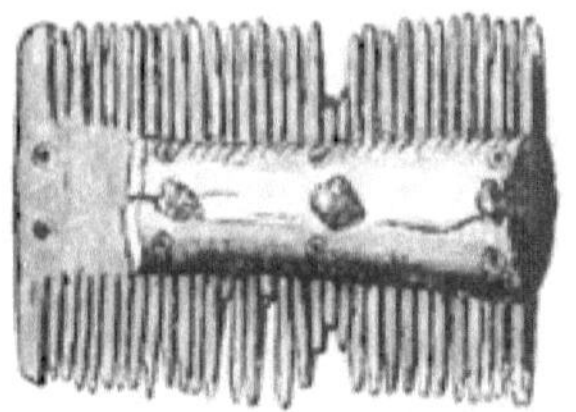

Fig. 256.—Bone Comb from the
Knowe of Saverough, Orkney (½).

consider that the neighbouring Celtic races in Scotland and
Ireland were in the habit of erecting similar island abodes,
while there is not a particle of evidence in favour of the idea
that such structures originated with the Roman conquerors
of Britain or its Saxon invaders.

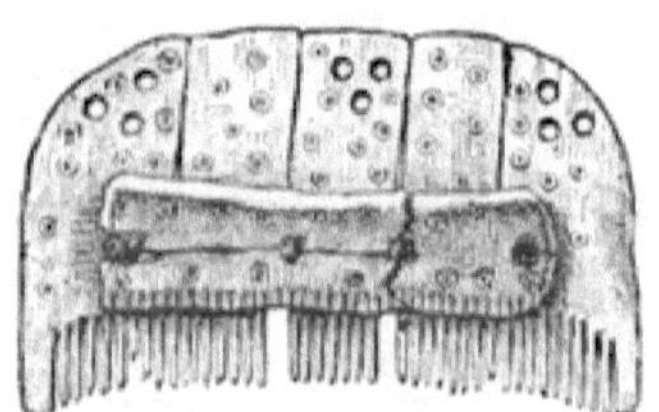

Fig. 257.—Comb of Bone found in the
Broch of Burrian, Orkney (½).

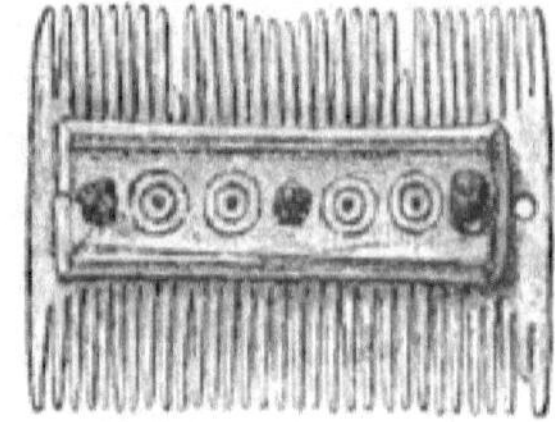

Fig. 258.—Bone Comb from the
Broch of Burrian, Orkney (½).

The resemblance between the remains found in the
Scottish and Irish lake-dwellings, as well as in many other
antiquarian finds of Celtic character, must also not be overlooked. Combs, similar in structure and ornamentation to

those from Buston, have been found in several of the Irish crannogs (Fig. 255);[1] in the Brochs and other antiquities of the north of Scotland, as at the Knowe of Saverough, Orkney (Fig. 256); the Broch of Burrian (Figs. 257 and 258); and in many of the ruins of the Romano-British towns, as at York and Uriconium (Fig. 259).[2] Iron knives and shears, variegated beads of impure glass, with grooves and spiral marks, ornaments of jet and bronze, implements of stone, bone, and horn, besides querns, whetstones, etc., are all common to Celtic antiquities wherever found.

Fig. 259.—Bone Comb from Uriconium (⅔).

Canoes are so invariably found associated with crannogs, that their discovery in lakes and bogs has been considered

[1] " It (the comb, Fig. 255) is 2½ inches long, and 1¾ deep, and the three pectinated portions are held together by flat sides, decorated with scrolls and circles. The top or handle shows a triple open-work decoration, and the side-pieces are grooved at one end for receiving the clasp of a metal tooth, which replaced one of the lost bone ones."

Double-edged combs " vary from 3 to 4½ inches in length, and from 1¼ to 2¼ across, the teeth portions being double, and passing through and through the sides to which they were riveted."

" The crannogs of Dunshaughlin, Ardakillen, and Cloonfinlough, and the street cuttings in the city of Dublin, have afforded nearly all the specimens of which the localities have been recorded. The total number of combs at present in the collection, including those on the ' Find Trays,' is eighty. Many of these combs are but fragmentary ; yet in each a sufficiency has been preserved to enable us to judge of the original size, and also of its style of ornamentation, which generally consists of transverse or oblique grooves, diced-work, interlacings, dotted lines, and circles surrounding a central indented spot."—(Wilde's *Catalogue*.)

[2] " The comb drawn in the sketch (Fig. 259) is an elaborate affair, consisting of central piece or pieces, with teeth ; the upper portion being strengthened in front and behind by transverse pieces of ornamented bone, riveted together by now much corroded iron rivets ; three of the rivets are reproduced."—(*Roman City of Uriconium*, p. 86.)

by Dr. Stuart as an indication of the existence of the latter. This may be true in some cases, but in others, such as Closeburn, Lochwinnoch, and Loch Doon, three of the examples cited by him, it is more probable that the canoes were used by the occupiers of the mediæval castles in the vicinity of which they were found. From these and many other instances that have come under my notice, I have come to the conclusion that dug-out canoes do not indicate such great antiquity as is commonly attributed to them, nor do they therefore necessarily carry us back to prehistoric times.

While some fragments of the pottery collected on the Ayrshire crannogs (all of which include Samian ware) are undoubtedly Romano-British, others as certainly point to a different period and source. I am informed on the best authority that all the portions showing any appearance of glaze, such as those represented by Figs. 181 to 184, were manufactured in mediæval times; but on the other hand that others (Fig. 186) may belong to the same class of fictile ware as was used for mortuary purposes in Pagan times. My own knowledge of the subject is too slender to guide me in forming an opinion on these points; but, considering how little is actually known of the pottery of the transitional period between that of the sepulchral urns and mediæval times, any conflicting inferences that may be deduced from such a line of discussion need not, at least in the meantime, interfere with the chronological conclusions pointed at elsewhere. The statement made by Grose (see page 155), that the monks of the monastery at Friars' Carse used to lodge their valuable effects on the artificial island in the loch when the English made inroads into Strathnith, suggests a possible, and perhaps probable, escape from any difficulty of this sort.

Again, while it is evident from the parallel striation on

some of the fragments, on which the traces of glaze still remain, that they were manufactured on the wheel, an inspection of Figs. 160, 161, and 181, proves that hand-wrought vessels were also manufactured after the introduction of glaze. That sepulchral urns were all made by hand is not a sufficient proof that the wheel was not simultaneously in use among the same people. The extraordinary conservatism, as regards changes in religious ceremonies, displayed by all nations and in all ages, is a sufficient explanation of the persistency with which the hand-wrought vessels have continued in use for sepulchral purposes. Hence, I can conceive that to substitute for the latter others made on the wheel, though at the same time largely used for domestic purposes, would have been an intolerable innovation on the religious customs of the people.

The Rev. Canon Greenwell, F.R.S., in discussing the question whether the various early sepulchral vessels were especially made for the purposes of burial, or were originally manufactured for domestic use, thus writes :—

"But perhaps the strongest objection to their having fulfilled a purpose in the household, is the fact that they possess but little in common with the pottery which, without much doubt, is domestic. It is true that not very much of this has been discovered, but quite enough has been found to enable us to judge pretty accurately of its character. It has not, indeed, been proved conclusively that the people who occupied the hut-circles and pit-dwellings were those who erected the barrows so often met with in close proximity to them ; but if we may judge, as I think we fairly may, from the identity of the flint implements found in each, there can be little doubt that they were, the one the dwelling-place, the other the burial-place, of the same people. Now, the pottery which has been discovered on the site of dwelling-places is a dark-coloured, hard-baked, perfectly plain ware, without ornament of any kind, is in fact just what we would expect domestic pottery to be, and has nothing in which it resembles the sepulchral vessels. And more than this, so far as I know of my own experience, or can learn from that of others, no whole vessel, or even fragments, **of** the ordinary sepulchral pottery of the barrows or other places of sepulture has ever been met with **in** connection with places of habitation."—(*British Barrows*, p. **106.**)

That many of these relics were the **products** of a refined **civilisation, is not more** remarkable **than the** unexpected and strangely discordant circumstances **in** which they **have been** found. For this reason it might be supposed that the crannogs were the headquarters of thieves and robbers, where **the proceeds** of their marauding excursions among the surrounding Roman provincials were stored up. The inferences derived **from a** careful consideration of all the facts do **not appear to me** to support this view, nor do they uphold **another view, sometimes** propounded, viz., that they were **fortified islands occupied** by the guardian soldiers of the **people. Indeed, amongst** the relics military remains **are only feebly** represented by a few iron daggers and spear-**heads, one or two** doubtful arrow-points, and **a** quantity of **round pebbles and** so-called slingstones. On the other hand, **a very large percentage of the** articles consists of querns, hammer-stones, **polishers,** flint-flakes, and scrapers; stone **and** clay spindle-whorls, pins, needles, and bodkins; knife-**handles of** red-deer horn, together with many other **imple-ments of the same** material; bowls, ladles, and **other** vessels **of wood, some of which were turned on** the **lathe;** knives, axes, **saws,** hammers, chisels, and gouges **of iron;** several **crucibles, lumps of iron** slag, and **other remains of** metals, etc. **From all** these, **not to** mention **the great** variety of **ornaments,** there can **be no** ambiguity as **to the** testimony **they afford** of **the** peaceful prosecution **of** various arts and **industries by** the lake-dwellers.

Proofs of a prolonged **but** occasionally interrupted occupancy are **also** manifested by the great accumulation of débris over the wooden pavements, the size and contents of the kitchen-middens, and the superimposed hearths so well observed at Lochlee.

From the respective reports of Professors Owen, Rolleston, and Cleland, on a selection of osseous remains taken from the lake-dwellings at Dowalton, Lochlee, and Buston (see pp. 50, 139-143, 236-239), we can form a fair idea of the food of the occupiers. The Celtic short-horn (*Bos longifrons*), the so-called goat-horned sheep (*Ovis aries*, variety *brachyura*), and a domestic breed of pigs, were largely consumed. The horse was only scantily used. The number of bones and horns of the red-deer and roe-buck showed that venison was by no means a rare addition to the list of their dietary. Among birds, only the goose has been identified, but this is no criterion of the extent of their encroachment on the feathered tribe, as only the larger bones were collected and reported upon. To this bill of fare the occupiers of Lochspouts crannog, being comparatively near the sea, added several kinds of shell-fish. In all the lake-dwellings that have come under my own observation, the broken shells of hazel-nuts were in profuse abundance.

From the number of querns and the great preponderance of the bones of domestic over those of wild animals, it may be inferred that, for subsistence, they depended more on the cultivation of the soil and the rearing of cattle, sheep, and pigs, than on the produce of the chase.

There is, in my opinion, only one hypothesis that can satisfactorily account for all the facts and phenomena here adduced, viz., that the lake-dwellings in the south-west of Scotland were constructed by the Celtic inhabitants as a means of protecting their lives and movable property when, upon the frequent withdrawal of the Roman soldiers from the district, they were left, single-handed, to contend against the Angles on the east, and the Picts and Scots on the north. It is not likely that these rich provincials, so long accustomed to the comforts and luxury of Roman civilisation, or their descendants in the subsequent kingdom of Strathclyde, would become the assailants of such fierce and lawless enemies, from whom, even if conquered, they could derive no benefit. Hence their military tactics and operations would assume more the character of defence than aggression, and in order to defeat the object of the frequent and sudden inroads of the northern tribes, which was to plunder the inhabitants rather than to conquer the country, experience taught them the necessity of being prepared for emergencies by having certain places of more than ordinary security where they could deposit their wealth, or to which they could retire as a last resource when hard pressed. These retreats might be caves, fortified camps, or inaccessible islands, but in localities where no such natural strongholds existed, the military genius of the Celtic inhabitants, prompted perhaps by inherited notions, led them to construct these wooden islands. Since the final departure of the Romans till the conquest of the kingdom of Strathclyde by the Northumbrian Angles, a period of several centuries, this unfortunate people had few intervals of peace (see pp. 249 to 259), and, with their complete subjugation, ended the special function of the Lake-Dwellings as a national system of protection. No doubt some of them, as well as caves and such hiding-places, would continue to afford a refuge to straggling remnants of natives rendered desperate by the relentless persecution of their enemies, but, ultimately, all of them would fall into the hands of the Saxon conquerors, when, henceforth, they would be allowed to subside into mud or crumble into decay.

Amongst extraneous antiquarian discoveries with which we may profitably compare the remains from the lake-dwellings, there are two remarkable groups recently brought to light by the exploration of two caves, both of which, singularly enough, are within the confines of the ancient kingdom of Strathclyde. These are a collection of relics from the Victoria Cave near Settle, Yorkshire,[1] and another from the Borness Cave, Kirkcudbrightshire.[2] Though the objects indicative of high-class art found in the latter are not so numerous as those from the former, yet they are, as a whole, of a similar character and type. They consist of fragments of Samian ware, bronze fibulæ, and other ornaments of bronze and horn, spindle-whorls, and a large quantity of implements and weapons made of stone, bone, and horn, etc., all of which bear a striking resemblance to the corresponding objects from the Lake-Dwellings. It would exceed my limits to enter upon a minute and critical comparison of the important results obtained from these two independent sources, viz., the Caves and the Lake-Dwellings ; I cannot, however, resist quoting the following remarks by Professor Boyd Dawkins, regarding the date of habitation of the Victoria Cave, which might, with equal appropriateness, be applied to the latter :—

"There can be no doubt but that this strange collection of objects was formed during the sojourn of a family for some length of time in the cave ; we have to account for the presence of so many articles of luxury in so strange and wild a place. The personal ornaments, and the Samian ware, are such as would have graced the villa of a wealthy Roman, rather than the abode of men who lived by choice in recesses in the rock. In the coins we have a key which explains the difficulty. Some belonged to Trajan and Constantine, others to Tetricus (A.D. 267-273), while others are barbarous imitations of Roman coins, which are assigned by numismatists to the period just about the time of the Roman evacuation of Britain. These objects, therefore, could not have been introduced into the cave before the end of the fourth century, or just that time when the historical

---

[1] *Collect. Antiq. and Journal of Anthrop. Inst.* vol. i.

[2] *Proceed. of Soc. of Antiq. Scot.* vol. x. p. 476; xi. p. 305 ; and xii. p. 669.

record shows us that the province of Roman Britain was suffering from the anarchy consequent on the withdrawal of the Roman troops. In the year 360, the savage Picts and Scots, pent up in the north by the Roman walls, broke in upon the unarmed and rich provincials, and carried fire and sword as far south as London. Their ravages were repeated from time to time, until the Northumbrian Angles finally conquered the Celtic kingdom of Strathclyde. It must nevertheless be admitted that, so long as the Celts of Strathclyde held their ground against the Angles, they would certainly follow the mode of life and the manners and customs handed down to them by their forefathers, the Roman provincials. And therefore, it is very probable that these objects of Roman culture may have been used in that district which was the Northumbrian border, long after they had ceased to be used in the regions conquered by the English. To say the least, there are two extremes between which the date must lie—the fourth and fifth centuries, as shown by the barbaric coins, and the year 756, when Eadberht finally conquered Strathclyde. It cannot be later, because of the presence of Roman, and the absence of all English cultus. The cave, situated in a lonely spot, and surrounded by the gnarled and tangled growth of stunted yews, oaks, and hazel, which still survive in one or two places in the neighbourhood, as samples of the primeval forest, would afford that shelter from an invader of which a native would certainly take advantage. We can hardly doubt that it was used by unfortunate provincials who fled from their homes, with some of their cattle and other property, and were compelled to exchange the luxuries of civilised life for a hard struggle for common necessaries. In no other way can the association of works of art of a very high order with rude and rough instruments of daily use be accounted for. In that respect, therefore, the Victoria Cave affords as true and vivid a picture of the troublous times of the fourth and fifth centuries as the innumerable burned Roman villas and cities; in the one case, you get a place of refuge to which the provincials fled; and in the other, their homes which have been ruthlessly destroyed."— (*Journal of Anthrop. Inst.* vol. i. p. 64.)

The presence of a Saxon coin in one of the Ayrshire examples is in no way inconsistent with the general views here advocated, as we have undoubted evidence that the country had been occupied, at least temporarily, by the Saxons as early as the date to which Mr. Evans assigns this coin.

Turning now to the Celtic area beyond the limits of the Scottish portion of the kingdom of Strathclyde, I may at once state that there are no data derived from an examination of its artificial islands, nor any relics of their occupiers, which can give even an approximate notion of their chronological range.

In localities where the Celtic races were not supplanted by foreigners, it would be strange indeed, and altogether at variance with archæological experience, as propounded by the learned author of *The Past in the Present*,[1] if the habit of resorting to isolated and inaccessible islands for safety would be all at once abandoned whenever the greater security afforded by stone buildings became known. Hence, in the Irish annals we find frequent mention made of crannogs down even to the middle of the seventeenth century, and Dr. Joseph Robertson has quoted several historical passages to prove that certain crannogs in Scotland—as, for example, those of the Loch of Forfar, Lochindorb, Loch Canmor, and Loch-an-eilan, etc.—survived to the middle ages. Many of these, however, were strong mediæval castles built for a different purpose, and had nothing in common with the crannogs proper beyond the fact of their insular situation.

From an etymological analysis of the earliest topographical nomenclature of Britain, as well as from other sources, there is abundant proof that in former times a Celtic population occupied nearly the whole of the island. Ultimately, however, these Celts were driven by successive waves of immigrants to the far north and west, and it becomes an important inquiry to determine if, in these localities from which they were expelled, there still exist any

---

[1] *The Past in the Present*, by Arthur Mitchell, LL.D.

traces of lake-dwellings. I have already remarked that no remains of wooden islands have been found in the south-east of Scotland. This, however, may be due to want of research on the part of antiquaries, and other causes, which so long kept us altogether in the dark regarding the phenomena of lake-dwellings in this country; and, indeed, some curious indications have already been supplied by independent observers as to their existence, not only in the south-east of Scotland, as at Balgone in East Lothian (see footnote, page 249), but in several parts of England and Wales.

CHAPTER VI.

SUPPLEMENTARY.

REMAINS OF LAKE-DWELLINGS IN ENGLAND.

As indicated by the title, the special object and scope of this work is to illustrate the phenomena of lake-dwellings as explored in Scotland. This limitation has been adopted, not with the intention of implying that there is any necessary identity between the area marked out by the general distribution of lake-dwellings and that included within the geographical limits of the kingdom of Scotland, but because, hitherto, their recorded remains south of the Scottish border were so few and undecided in character that there could hardly be any justification in deviating from the commonly entertained opinion that these structures were not to be found in England. But after finishing my labours under this impression, some additional facts have come under my cognisance which greatly strengthen the idea, rather hesitatingly expressed at the conclusion of my last chapter, viz., the probability of the lake-dwelling district being found to coincide with the former extension of the Celtic area in Britain. Partly to support this theory, but more particularly to make this work more complete by including the actual materials that could be supplied were it to appear under the more comprehensive title which the substitution of the word *British* instead of *Scottish* would give to it, I have collected together, in the form of this supplementary chapter, all the scattered notices of such trustworthy observations as can now be fairly construed to indicate the sites of lacustrine abodes in England. It will be noticed that some of the recorded observations here reproduced were actually made before antiquaries had time to realise the magnitude of the Continental lake-dwellings, or the subsequent promulgation of Dr. Robertson's views on the Scottish crannogs, and consequently at a time when their real importance was apt to be overlooked; otherwise, it is impossible to conceive how such highly suggestive facts did not at once lead to more definite information.

Wretham Mere, Norfolk.

Sir Charles J. F. Bunbury, as early as 1856, noticed some appearances in a drained *mere* near Wretham Hall, Norfolk, which clearly point to being the remains of a lake-dwelling. In a communication on the subject to the Geological Society,[1] he says :—" About Wretham there are several *meres*, or small natural sheets of water, without any outlet. The one to which my attention was particularly called by Mr. Birch occupied about forty-eight acres, and was situated in a slight natural depression, the ground sloping gently to it from all sides. The water has been drawn off by machinery, for the purpose of making use, as manure, of the black peaty mud which formed the bottom. . . .

" Numerous horns of red-deer have been found in the peaty mud, generally (as I was informed) at 5 or 6 feet below the surface, seldom deeper; many attached to the skull, others separate, and with the appearance of having been shed naturally. What is most remarkable, several of those which were found with the skulls attached had been

---

[1] *Quarterly Journal of the Geological Society*, vol. xii. p. 255, May 7, 1856.

sawn off just above the brow antlers—not broken, but cut off clean and smoothly, evidently by human agency. Some of these horns are of large size, measuring 9 inches round immediately below the brow antler. . . .

"Numerous posts of oak-wood, shaped and pointed by human art, were found standing erect, entirely buried in the peat."

Pile Structures at London Wall.

On December 18th, 1866, General Lane Fox, F.S.A., read a paper at the Anthropological Society, entitled "A Description of certain Piles found near London Wall and Southwark, possibly the remains of Pile-Buildings."

The author commenced by observing that his attention was directed to this locality by a short paragraph in the *Times* of the 20th October 1866, stating that upwards of twenty cart-loads of bones had been dug out of the excavations which were being made for the foundations of a wool warehouse near London Wall.

The excavation in question commenced at 40 yards south of the street pavement; therefore, in all probability, at about 70 or 80 yards from the site of the old wall. The area excavated at the time of General Lane Fox's visit was of an irregular oblong form, 61 yards in length, running north and south, and 23 yards wide.

A section of the soil consisted of—

" 1. Gravel similar to Thames ballast at a depth of 17 feet towards the north, inclining to 22 feet towards the south end.

" 2. Above this, peat of unequal thickness, varying from 7 to 9 feet.

" 3. Modern remains of London earth, composed of the accumulated rubbish of the city."

Regarding the remains of piles, the author makes the following important observations :—

"Upon looking over the ground, my attention was at once attracted by a number of piles, the decayed tops of which appeared above the unexcavated portions of the peat, dotted here and there over the whole of the space cleared. I noted down the positions of all that were above ground at the time; and as the excavations continued during the last two months, I have marked from time to time the positions of all the others as they became exposed to view.

"Commencing on the south, a row of them ran north and south on the west side, to the right of these a curved row, as if forming part of a ring. Higher up and running obliquely across the ground was a row of piles, having a plank about an inch and a half thick and a foot broad placed along the south face, as if binding the piles together. To the left of these another row of piles ran east and west; to the north-east again were several circular clusters of piles; these were not in rings but grouped in clusters, and the piles were from eight to sixteen inches apart. To the left of this another row of piles and a plank two inches thick ran north and south. There were two other rows north of this and several detached piles, but no doubt several towards the north end had been removed before I arrived.

"The piles averaged six to eight inches square; others of smaller size measured four inches by three; and one or two were as much as a foot square. They appeared to be roughly cut, as if with an axe, and pointed square; there was no trace of iron shoeing on any of them, nor was there any appearance of metal fastenings in its planks; they may have been tied to the piles, but if so, the binding material had decayed. The grain of the wood was still visible in some of them, and they appear to be of oak. The planks averaged from one to two inches thick. The points of the piles were inserted from one to two feet in the gravel, and were, for the most part, well preserved, but all the tops had rotted off at about two feet above the gravel, which I conclude must have been the surface of the ground, or of the water at the time these structures were in existence."

The relics were exclusively found in the peat or middle layer (which varied from 7 to 9 feet in thickness), but "interspersed at different levels from top to bottom throughout it." According to the author the vast majority of them belonged to the Roman era. He says : " Amongst them are quantities of broken red Samian pottery, mostly plain, but some of it depicting men and animals in relief; one specimen is stamped with the name of Macrinus. All this pottery, in the opinion of Mr. Franks, to whom I showed it, is of foreign manufacture. Other samples are of the kind supposed to have been manufactured in the Upchurch Marshes in Kent, and upon the site of St. Paul's Churchyard. Bronze and copper pins, iron knives, iron and bronze stylus, tweezers, iron shears, a piece of polished metal mirror, so bright that you may see your face in it (this Dr. Percy has pronounced to be of iron pyrites, white sulphuret of iron without alloy), an iron double-edged hatchet, an iron implement, apparently for dressing leather, a piece of bronze vessel, and other bronze and iron implements, which, thanks to the preserving properties of the peat, are all in excellent preservation. Amongst these were also a quantity of leather soles of shoes or sandals, some apparently much worn, and others, being thickly studded with hobnails, may be recognised as the caliga of the Roman legions; also a piece of a tile with the letters P . PR . BR . stamped upon it. Specimens of these are on the table. The coins found are those of Nerva, Vespasian, Trajan, Adrian, and Antoninus Pius. . . .

" In addition to the Roman relics above mentioned, others of ruder construction remain to be described. They consist of what, in the absence of any evidence respecting their uses, may be called handles and points of bone. The former are composed of the metacarpal bones of the red-deer and *Bos longifrons* cut through in the middle, and roughly squared at the small end; the others, which are called by the workmen spear-heads, are pointed at one end and hollowed out at the other, as if to receive a shaft. Both Professor Owen and Mr. Blake concur in thinking these implements may possibly have been formed with flint, but I cannot ascertain that they were found at a lower level than the Roman remains, nor have any flint implements, to my knowledge, been found in the place. With them were also found the two bone skates on the table; they are of the metacarpal bone of a small horse or ass, one of which has been much used on the ice. Exactly similar skates also of the metacarpal of the horse or ass have been found in a tumulus of the stone period at Oosterend in Friesland; a drawing of them is given in Lindenschmit's Catalogue of the Museum at Mayence, etc. Others have also been found in Zeeland, at Utrecht, and in Guelderland, and there is a specimen in the Museum at Hanover. Professor Linden-schmit attributes all these to the stone period, but the speci-mens on the table are evidently of the iron age, the holes in the back having been formed for the insertion of an iron staple. Similar skates have been found in the Thames, but they have not hitherto been considered to date so early in England as in Roman times."

Throughout the peat were several kitchen-middens. One deposited a foot and a half above the gravel is thus described: "A layer of oyster and mussel shells about a foot thick, with a filtration of carbonate of lime permeating through the moss. In this kitchen-midden Roman pottery and a Roman caliga were found. Close by, the point of a pile, part of which is exhibited, was found upright in the peat; it had been driven in in such a manner that the point descends to the level of the kitchen-midden and no further. Now, as a pile, in order to obtain a holding, must have been driven at least two feet in the ground, it is evident the peat must have grown at least one foot above the summit of the kitchen-midden before this pile was driven in."

A second kitchen-midden is noted at a height of $3\frac{1}{2}$ feet above the gravel, " composed of oyster, cockle, and mussel shells, and periwinkles, with Roman pottery and bones of the goat and *Bos longifrons*, etc., split lengthwise as if to extract the marrow, with the skulls broken and the horns cut off. It is about a foot and a half thick in the centre, thinning out towards the ends as a heap of refuse would naturally do, and from 12 to 14 feet long; above this is peat for about a foot or a foot and a half, and above the peat another kitchen-midden of the same kind as the preceding. Lastly, the soles of shoes and Roman pottery of the same kind as that found lower down have been taken out at the very top of the peat."

The author being subsequently anxious to obtain further evidence as to the thickness of the stratum in which the Roman remains were found, states that he determined to watch the workmen for four or five hours together during several successive days, while they dug from top to bottom, commencing with the superficial earth, and passing through the peat to the gravel below. The result was as follows: " Roman red Samian ware is found as high as 13 feet from the surface, but very rarely, and in small quantities. At 15 feet it is frequently found, and from that depth it increases in quantity till the gravel is reached at 18 to 21 feet. The chief region of Roman remains is within 2 to 3 feet of the gravel."

Amongst the animal remains were, according to Professor Owen, those " of the horse or ass, the red deer, the wild boar, the wild goat (*bouquetin*), the dog, the *Bos longi-*

frons, and the roebuck. The horns of the roebuck, I afterwards ascertained, were all found at a higher level. These, and also the horse and goat, entered the superficial earth, in which glazed pottery was also found; but the remainder, including the red deer, wild boar, and *Bos longifrons*, appeared, so far as my observations enabled me to judge, to be confined to the peat."

Subsequently Mr. Carter Blake identified amongst these remains no less than four different kinds of the genus *Bos*, viz., *primigenius*, *trochoceros*, *longifrons*, and *frontosus*; as also a specimen of the ibex of the Pyrenees.

Some human skulls were also found in the lowest formation of the peat, or immediately over the gravel. Along with the skulls only three human bones were found; but this, according to the author, was not the result of an oversight, as both the Celts and the Romans were known to have practised decapitation.

The piles at the south end were identified as elm, the remainder were oak (*Quercus robur*).

General Lane Fox stated that recently similar piles with large horizontal beams and Roman pottery were discovered in New Southwark Street.

I find it impossible, even with the above large extracts, to give more than a very general idea of this most interesting and highly suggestive paper, and the important discussion to which it gave rise in the Society.

Crannog in Llangorse Lake, near Brecon, South Wales.

In Keller's book on Lake-Dwellings,[1] there is a notice of a "crannoge, or stockaded island, in Llangorse Lake, near Brecon (South Wales)," by the Rev. E. Dumbleton, M.A., in which the author describes an island 90 yards in circumfer-

---

[1] *Lake-Dwellings*, by Keller (2d ed., trans. by Lee), p. 660.

ence, the highest part of which is 5 feet above the level of the water, on which "some small trees and brushwood have fastened," and around which numerous cleft oak-beams have been detected. In examining the interior by perpendicular openings, they invariably led down to the shell-marl, "showing first a stratum of large, loose stones, with vegetable mould and sand, next (about 18 inches above the marl) peat, black and compact; and beneath this, the remains of reeds and small wood. This fagot-like wood presented itself abundantly all round the edges of the island, and in the same relative position, namely, immediately upon the soft marl; the object of it being, of course, to save the stones from sinking." Pieces of charcoal, broken bones, "a piece of leather pierced with several holes, in some of which, when discovered, the remains of a thong might be observed," three or four scraps of pottery, and a stone that seemed to have been ground, are the only indications of human occupancy recorded. Remains of log platforms, which were observed, are also described in this article. Some of the bones were sent to Professor Rolleston, of Oxford, who wrote that "the chief points of interest respecting these were: first, the presence of two varieties of horse—one small, such as a Welsh pony is; and the other large (as I am informed large horses appear to have existed, as well as mere Galloways, in the very earliest human periods in this country); and, secondly, the smallness of the then ordinarily eaten mammals, *Sus*, *Bos*, *Ovis*. The horse was eaten formerly, especially by the Pagans, and it may have been eaten by the inhabitants of your crannoge; but there is no evidence, from splitting or burning, that they did so." "Some other bones, found subsequently, were exhibited at the meeting of the British Association at Exeter, and were examined by Mr. W. Boyd Dawkins, who pronounced them to be those of the red deer, the wild boar, and the *Bos longifrons*. He stated that the group altogether, from the greater percentage of wild than domestic animals, indicated a remote period."

Barton Mere, near Bury St. Edmunds.

Professor Boyd Dawkins, under the heading *Habitations in Britain in the Bronze Age*, writes as follows :—" Sometimes, for the sake of protection, houses were built upon piles driven into a morass or bottom of a lake, as for example in Barton Mere (explored by Rev. Harry Jones in 1867,—*Suffolk Inst. of Archæology and Natural History*, June 1869), near Bury St. Edmunds, where bronze spear-heads have been discovered, one 18 inches long, in and around piles and large blocks of stone, as in some of the lakes in Switzerland. Along with them were vast quantities of the broken bones of the stag, roe, wild boar, and hare, to which must also be added the urus, an animal proved to be wild by its large bones, with strongly-marked ridges for the attachment of muscles. The inhabitants also fed upon domestic animals— the horse, short-horned ox, and domestic hog, and in all probability the dog, the bones of the last-named animal being in the same fractured state as those of the rest. Fragments of pottery were also found. The accumulation may be inferred to belong to the late rather than the early Bronze Age, from the discovery of a socketed spear-head. This discovery is of considerable zoological value, since it proves that the urus was living in Britain in a wild state as late as the Bronze Age. It must, however, have been very rare, since this is the only case of its occurrence at this period in Britain with which I am acquainted."—(*Early Man in Britain*, p. 352.)

Professor T. Rupert Jones on English Lake-Dwellings.

In 1878, Professor T. Rupert Jones, F.R.S., communicated to *Nature* a short notice on " English Lake-Dwellings and Pile Structures," in which, after drawing attention to the previously published articles of General Lane Fox and Sir Charles Bunbury, he writes as follows:—

"Since writing the above I have been informed that Mr. W. M. Wylie, F.S.A., referred to this fact in '*Archæologia*,' vol. xxxviii. in a note to his excellent memoir on lake-dwellings. I can add, however, that remains of *Cervus elaphus* (red deer), *C. dama?* (fallow-deer), *Ovis* (sheep), *Bos longifrons* (small ox), *Sus scrofa* (hog), and *Canis* (dog), were found here, according to information given me by the late C. B. Rose, F.G.S., of Swaffham, who also stated in a letter dated August 11th, 1856, that in adjoining meres, or sites of ancient meres, as at Saham, Towey, Carbrook, Old Buckenham, and Hargham, cervine remains have been met with; thus at Saham and Towey *Cervus elaphus* (red deer), at Buckenham *Bos* (ox) and *Cervus capreolus* (roebuck); at Hargham *Cervus tarandus* (reindeer).

"The occurrence of flint implements and flakes in great numbers on the site of a drained lake between Sandhurst and Frimley, described by Captain C. Cooper King in the *Journal of the Anthropological Institute*, January 1873, p. 365, etc., points also in all probability to some kind of lake-dwelling, though timbers were not discovered.

"Lastly, the late Dr. S. Palmer, F.S.A., of Newbury, reported to the 'Wiltshire Archæological Society' in 1869 that oaken piles and planks had been dug out of boggy ground on Cold Ash Common, near Faircross Pond, not far from Hermitage, Berks."—(*Nature*, vol. xvii. p. 424.)

Holderness, York.

A few weeks ago my attention was directed by Mr. Joseph Anderson to a communication which he had just received from a gentleman near Bridlington anent some antiquarian remains indicating lake-dwellings in that district, of which, at my request, the discoverer has kindly favoured me with the following interesting notice :—

" ULROME GRANGE,
LOWTHORPE, HULL, Feb. 28, 1882.

" DEAR SIR,—This part of the county of York (Holderness) appears formerly to have been intersected by numerous irregular lakes, which were drained about eighty years ago.

" In the spring of the year 1880 the Commissioners of the Beverley and Barmston Drainage found it necessary to deepen one of these drains (the branch called the Skipsea drain).

" A short time after this was done I was walking in one of my fields adjoining, and picked up some perforated bone implements. I shortly afterwards had the earth, which had been excavated at this place, turned over, and found more implements of the same class. Also two made from the antlers of the red deer, and a small piece of red ochre, with several stones which bear traces of having been utilised.

" In the month of May 1881, the water in the drain at that time being very low, and having obtained the services of half a dozen men accustomed to similar work, I had the water dammed, and dug through the peat to a bed of gravel, 9 ft. 6 in. from the surface.

" We found three more perforated bone implements, all in the side of the drain, and at the depth of seven feet, also several stakes and piles with remains of brushwood.

" I then determined, when opportunity offered, to excavate in the field, and proceeded to do so in December last.

" We commenced by digging a trench parallel with the drain and sixty feet in length. This trench and the drain formed two sides of a square, running north and south. Commencing at the south end, we came upon a layer of gravel at the depth of two feet, which dips to the north, and is overlaid by a bed of peat, six feet in thickness, at the north end of the trench.

" As this trench filled with water, we began at the same point and dug a similar one on the south side, running east and west, and connected the first trench with the drain. The gravel slopes also to the west, and dips quite abruptly when at a distance of forty feet from the drain. When the trenches were dug a gravel slope at the south-east corner of the square prevented the water from running out of the first trench. I therefore had the earth turned over on this slope, when we found great numbers of stakes, with some brushwood, the earth being a peaty marl.

"When clear of the slope there is a decided layer of brushwood about two feet thick, also studded with stakes, and along the inner side of the south trench we found a number of piles from five inches to seven inches in diameter, in a line, and mostly upright. One of these we got out quite perfect. It is of oak wood, four feet in length, six inches in diameter, and has a forked top which has apparently been intended for carrying a horizontal beam or support. The piles are about four feet apart. One had given way and been replaced.

"As the trench is not exactly in a line with the piles, several are now left standing and partially exposed. In this portion of the digging, we found several bones of animals, a peculiar grinding-stone of whinstone or granite, almost semicircular in shape, 12 inches long by 7 broad, a flint core, a stone with the centre hollowed, a pounding or hammer stone, and two fragments of rude pottery, evidently British.

"Hazel nuts are numerous; several I have picked out appear to have been opened by squirrels.

"After making these discoveries I suspended work, as I felt that I should like some one acquainted with similar explorations to give an opinion with respect to the course I ought to adopt.

"Whether the place is a lake-dwelling or not, further research will determine. It is undoubtedly a pile structure, and of a very early date.

"At this season the spring-tides tend to impede further investigation, the water having risen to the height of 7 feet in the trenches on the 19th inst. And as we may hope for warmer weather with longer days, I shall probably defer further exploration until April. I believe I have discovered another similar place, but on a larger scale, and the timbers appear much larger. The two are not more than half a mile apart, and are situated on the same lake as the earthworks and mound at Skipsea (described in Poulson's *Holderness*). In the meantime, any suggestions you may favour me with will be gladly received by yours very faithfully,

" THOMAS BOYNTON.

"Dr. Munro, *Kilmarnock.*"

Concluding Remarks.

It may be some time yet before further research will throw much additional light on the appearances and discoveries above recorded, but should they turn out to be the genuine remains of ancient lake habitations, it is more than probable that they will be found to be no exceptional instances, but remnants of a more widely distributed custom. Meantime, however, they appear to me sufficiently suggestive, especially when taken together with the evidence I have already produced as to the prevalence of such structures amongst the Celtic races in Scotland and Ireland, and the distinct statement made by Julius Cæsar that the Britons made use of wooden piles and marshes in their mode of entrenchment (*sylvis paludibusque munitum*), to entertain the hypothesis that the original British Celts, from whom in all probability have descended the modern Gaels, were an offshoot of the founders of the Swiss Lake-Dwellings, that they emigrated to Britain when these lacustrine abodes were in full vogue, and that, as they spread northwards and westwards over Scotland and Ireland, they retained, and probably practised, the habit of resorting to insular protection long after the custom had fallen into desuetude in Europe. As, however, the lake-dwelling mania subsided and gradually came to a close on the Continent, subsequent immigrants into Britain, such as the Belgæ, Angles, etc., being no longer acquainted with the subject, cultivated new principles of defensive warfare, or, at any rate, ceased to resort to the protection afforded by the artificial construction of lake-dwellings, whilst the first Celtic invaders, still imbued with their primary civilisation, when harassed by enemies and obliged to act on the defensive, had recourse to their peculiar and inherited system of protection, with such variations and improvements as better implements and the topographical requirements of the country suggested to them. Hence it would follow that the range of the British Lake-Dwellings, both in space and time, would vary according to circumstances and the vicissitudes of their founders; but, speaking generally, it is only reasonable to suppose that its limitation first commenced in those districts most accessible to fresh swarms of Continental immigrants. But this problem, as well as many other subsidiary questions which follow in the same line of inquiry, must be solved by further researches; and should these remarks in any way lead to renewed application in this department, they will serve a good end, whatever may be the result of the hypothesis thus broached regarding the primary sources of the ideas that led to the development of British Lake-Dwellings.

APPENDIX.

ADDITIONAL DISCOVERIES ON THE CRANNOG IN
LOCHSPOUTS.

As mentioned at page 182, the selection of the natural basin of
Lochspouts as the most suitable site for a reservoir for supplying
the town of Maybole with water had been announced shortly
after the excavation of the crannog (as far as was then possible
without an expensive cutting to reduce the level of the lake) had
been completed. In the course of the subsequent negotiations
with the proprietor and his agents, which ended in the final
adoption of this scheme, we have another proof of the interest
taken by Sir James Fergusson, Bart., in these antiquarian researches. The following extract, taken from the contractor's
specification for the work to be done within this lake-basin, pre-
paratory to its conversion into the proposed reservoir, requires
no explanation :—" After the water in the present loch has been
lowered, the bottom of the reservoir, to the extent to be pointed
out, to be excavated to a depth of about 3 feet, or to such further
depth as the engineer may consider it necessary, to remove the
peat and other matters. At the site of the supposed lake-
dwelling, the excavations to be so conducted that the structure
of the dwelling may be left entire, until such time as it is
thoroughly explored by a member or members of the Archæological Society of Ayrshire and Wigtownshire, or such person
appointed by them or by Sir James Fergusson, Bart., of Kilkerran, to see this exploration carried out. Any relics that may

U

be found during the excavation to be at once delivered to the party appointed to superintend the exploration, or to such other person as may be in charge in his absence." Just as the proof-sheets of my last chapter had come to hand, I received a note from Mr. William Henderson, C.E., engineer to the Maybole water-works, stating that the outlet at Lochspouts had been cut about 3 feet deeper, that the water was being drained off, and that the contractor was ready to begin the excavations on the site of the lake-dwelling. In anticipation that the result of these operations would furnish a satisfactory solution of some of the problems left undetermined in my previous report, the publication of this volume has been delayed for a few days in order to secure the desired information, and hence I am enabled to give the following short report of the additional discoveries made on this lake-dwelling.

On the receipt of Mr. Henderson's letter, I lost no time in making an appointment to meet him at Lochspouts, where I became more fully acquainted with the nature and extent of the proposed excavations. The débris formerly wheeled from the mound lay in two heaps just beyond the margin of the artificial island, but still within the boundaries of the reservoir. These, therefore, together with a complete section of the island, about 3 feet in thickness, were to be removed entirely beyond the rocky barrier. I understood that in clearing away the contents of this section, the wood-work, especially towards the margin of the crannog, and about the surrounding piles, was to have been left intact for some time, but when I revisited the scene of the operations a few days afterwards, I found that a gang of some forty or fifty men had made such progress that the whole section was completely removed, leaving nothing but small pillars here and there for the purpose of calculating the number of cubic yards excavated. All the horizontal beams and other wood-work were taken away, and nothing left above the base of the section except a few of the encircling uprights on the shore side of the crannog. My regret at this unexpected rapidity of the process of demolition was however considerably allayed when I found that Mr. James Mathewson, the inspector of the works, under whose vigilant eye the operations were conducted, had taken a most intelligent interest in the archæological phase of the remains, and had even taken notes of some of the phenomena which appeared to him most important. It is therefore to him I am chiefly indebted for the following details.

During the former explorations, the conjecture that the paved habitable surface, with its remains of hearths, relics, etc., then reached, was a secondary one superimposed upon the débris of a former habitation, was supported by the following observations, which could not, however, be verified by deeper excavations, owing to the rushing up of water :—

    (1.) The level of the log-pavement was considerably higher than the tops of the uprights forming the surrounding circles.

    (2.) In various places, when attempting to dig beneath it, ashes, charcoal, bones, hazel-nuts, and sea-shells were turned up. (See page 164.)

The evidence now produced left no doubt that this conjecture was well founded.

On the bank I was shown two heaps of oak beams which had been removed from the excavated débris, and amongst them were some of the ordinary transverses, containing square-cut holes at their extremities. Upon inquiry, I found that some of these, when exposed, were in position in the line of the surrounding stockade, with uprights projecting through the holes. One thick beam was deeply grooved, and resembled one found at Lochlee, figured and described at page 84. A few large flat planks, having a round handle-like projection at one end, some 18 inches long, had only one square-cut hole, sometimes close to this handle, and at other times at the opposite extremity. Another stout oak beam, 6 feet long, contained a series of round holes about an inch in diameter, and from five to six inches apart. The holes, which were on the broad side of the beam, were about 2 inches in depth, but only penetrated half through it, and from one of them portion of a wooden pin was extracted.

This beam was in a fragmentary condition, being, like many others, partially charred.

On examining the surface of the island, as now exposed, I noticed some very large oak beams, prepared like railway sleepers, and in one place, near the centre, there were some stones and clay as if they had formed the base of a fireplace; but the whole area was so muddy that it was difficult to say whether or not this was the exact surface of a former log-pavement.

On looking at the isolated pillars left standing, we noticed that their substance, which consisted of vegetable débris, mixed with brushwood, ashes, and in one place layers of clay, had a more or less stratified arrangement. The depth of the layer removed varied from $2\frac{1}{2}$ to 3 feet, and it appeared to me as if the island had sunk less towards its shore-side than on the far-off side, as the tops of the surrounding piles had become barely visible on the latter, whereas, on the former, not only were the piles exposed for about 18 inches or 2 feet, but some of the transverses were actually found in position lying over them.

At the junction of the gangway and island, a full view of which we now had, the uprights of both structures appeared to be on the same level, but as those of the former approached the shore, they became gradually more elevated, till, as mentioned at page 166, they projected above the grass.

As regards the deeper structures of the island, I was always of opinion, considering the amount of subsidence of its surface that had taken place, that their depth would be correspondingly great. This opinion was now shared by the engineer, contractor, and others, who judged more from the great solidity and firmness presented by the whole mass. In attempting to ascertain some further particulars by digging a hole in its centre, Mr. Mathewson writes as follows :—

" LOCHSPOUTS, 2d May 1882.

" DEAR SIR,—I have been instructed by the engineer to forward to you, by Wednesday at latest, any information gained by the sinking of the central shaft in the crannog.

" The mode adopted was to open a place about 12 feet square.

" The pump forwarded was only 3 inches diameter, and it was found that three men bailing with buckets were required to enable other two men to dig.

"A large mortised oak beam was found about 18 inches below present surface; still further down a few oak beams were lifted with broken portions of transverse (soft wood) beams adhering to under surface of the oak. This was at a depth of 3 feet 6 inches. A large flat stone, near to which was a compressed mass of grass, some ferns (common bracken), and fragments of moss, was also turned up.—I am, yours faithfully,

"JAMES MATHEWSON.

"*P.S.*—5.20 P.M. Men leaving. Found mortised beam (oak) with pin in hole. Beams as far as we can plunge a rod— $3\frac{1}{2}$ feet deep."

Writing subsequently, May 11th, Mr. Mathewson says:— "The sinking of the shaft was a failure through want of depth at outlet. Oak beams with cross layers of softer wood and brushwood were found all over the bottom of shaft. Some small jaw-bones were brought up from a depth of 2 feet 6 inches below present surface, as also some compressed ferns and grasses, a small quantity of ashes, and a trace of whitish clay. On Friday evening I turned over some of the formerly unmoved oak beams at a corner of the shaft, put the spade a foot further down and turned up a sandstone which had been used as a whetstone. It was irregularly shaped, 7 inches long and $2\frac{3}{4}$ broad. One flat face and a sloping edge were ground smooth by whetting. It was 1 foot 6 inches below present floor. In the near surroundings of the spot I also found ashes and traces of tough whitish clay and a few bones."

Again, writing on the 16th May, Mr. Mathewson says:— "I sounded shaft to-day, and found hard beams from 3 feet 9 inches to 4 feet 3 inches below present level of excavations. The shaft is rudely 3 feet 6 inches deep. In some crevices the iron bar went down to 6 feet from top of shaft, and again struck wood."

RELICS FOUND BELOW UPPER LOG-PAVEMENT.

But the chief evidence that the section now removed from off the island represented the débris of a former habitation, is derived from the relics found among its contents, which are as follows :—

(1.) *Whetstones.*—Three of these implements, the most modern-like that I have yet seen, were found to the west of the junction of the gangway with the crannog, and at a depth of 2 feet 6 inches. One is rectangularly-shaped and beautifully polished on all sides. It is made of a hard dark stone, and measures $7\frac{1}{4} \times 1\frac{3}{8} \times \frac{7}{8}$ inches; another is a smooth slightly oval-shaped rod, $5\frac{1}{2}$ inches long and about $\frac{3}{4}$ inch in diameter; the other is about the same length, but of a roughly quadrangular shape.

(2.) *Wooden Implements.*—A semi-globular piece of soft wood, 7 inches in diameter, and having a shallow cavity cut out of its flat surface, measuring 5 inches in diameter, and a uniform depth of $1\frac{1}{2}$ inch. Another cup-shaped vessel or implement, also of soft wood, was surrounded by a deep groove, across which were seen the remains of small wooden pins, some nine or ten in number, which penetrated through both its rims. The diameter of the central cup was $5\frac{1}{2}$ inches, and that of the whole vessel (including the outer rim, the groove, and the rim of the cup), $8\frac{1}{4}$ inches. A third article of wood consisted of a smooth flat beam of oak, 3 feet 6 inches long, 1 foot broad, and 4 inches thick, having a deep groove at one edge, and a stout pin-like projection from one end, as if it had other attachments. In the centre of this beam there was a round hole over which lay a handle-like elevation cut out of the solid, and having not only a vertical hole corresponding with the one in the lower portion, but also another passing horizontally through it, and immediately between the two former. This handle-like elevation was 2 feet 1 inch long, $4\frac{1}{2}$ inches broad, and rose into a slight arch in the middle, where the horizontal hole passed underneath, and in the line of continuation of the latter there was, on both sides, a slight hollow, as if worn out of the beam by friction. The whole was cut out of one piece of solid oak. These articles were found at a depth of about 2 feet below the former log pavement.

(3.) *Bronze Ornament.*—A double-spiral ornament of bronze wire, having six twists at one end and three at the other, was found at a depth of 1 foot 6 inches, and near the centre of the island. Its length is $1\frac{3}{4}$ inch (Fig. 260).

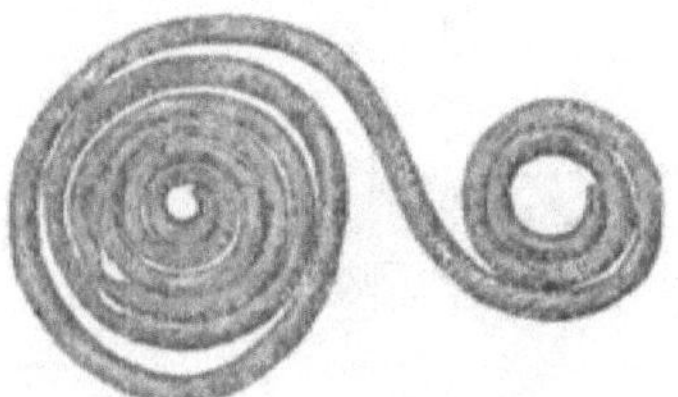

Fig. 260.—Bronze Ornament ($\frac{1}{1}$).

(4.) *Jet Ring.*—This article was found at a depth of $1\frac{1}{2}$ foot. Its diameter is $1\frac{1}{8}$ inch. The inside looks as if worn in one or two places by friction.

Besides the above, some hammer-stones, a quern, and two fragments of very rude pottery were found.

ARTICLES FOUND WHILE REMOVING THE STUFF FORMERLY WHEELED FROM OFF THE MOUND, *i.e.* ABOVE THE UPPER LOG-PAVEMENT.

(1.) *Rock Crystal.*—A conical piece of rock-crystal, evidently ground down to its present shape. The diameter of base is $\frac{1}{13}$ less than an inch, and the perpendicular height is $\frac{3}{4}$ of an inch. The base is not quite flat, but slightly convex, as will be seen from the annexed outline (Fig. 261). It scratches glass, but is scratched by a diamond, and depolarises a ray of light. Its specific gravity is 2·64.

Fig. 261.—Outline of Crystal Ornament ($\frac{1}{1}$).

(2.) *A Leaden Spindle-Whorl.*—A small bead-shaped portion of lead perforated with a round hole is supposed to be a spindle-whorl. Its diameter is ¾ of an inch.

(3.) *Bronze Ornament.*—This consists of a small semi-globular-shaped cup, ¾ of an inch in diameter, to which is attached a triangular-shaped handle-like projection, ¾ of an inch long (Fig. 262).

Fig. 262.—Bronze Ornament (½).   Fig. 263.—Amber-coloured Glass Bead (½).

(4.) *Glass Bead.*—This is a smooth, amber-coloured bead, variegated with a yellowish slag, and measuring ¾ inch in diameter, and $\frac{7}{10}$ of an inch deep (Fig. 263).

(5.) *Bronze Ring.*—A small slender ring of bronze, of the size of a finger-ring. It is penannular (but the ends are close, and might have been broken), and is ¾-inch in diameter.

(6.) *Jet Pendant.*—This is made of a circular piece of polished jet or cannel coal, rather less than 1½ inch in diameter, and ¼ inch thick, which is perforated by four quadrant-like spaces of uniform size and shape, so as to leave the form of a rectangular cross inscribed in a circle. The arms of the cross become a little broader as they approach the circumference, and on one surface they, as well as the circular portion, are ornamented by a row of incised circles, each circle having a small hollow in its centre. An incised line bounds each row of circles on both sides. All these incised lines, circles, and central hollows, were filled by a yellowish kind of enamel.

A little projection from the circle, opposite one of
the arms of the cross, is perforated transversely to
its surface by a small hole for suspension, but it is

Fig. 264.—Jet Ornament (½).

evident that previous to the making of this hole, it
was suspended by means of another hole, which
perforated it in an opposite direction, but from
which one side was broken off.   (Fig. 264.)

Dr. Joseph Anderson, to whom I sent this object for inspection, writes thus :—" I have nothing special to say of the jet
object sent to-day, except that it seems to be most certainly
Christian, and of an early Christian type.   It is the first jet
thing I have seen, having this Christian relationship, from any
of the early inhabited sites in Scotland.   The ornament is very
peculiar, and the form of the trinket most interesting, as it
compares with the form of the cross within a circle found on
the stones in Wigtownshire, though it has not the peculiar
appendage which marks the Chrisma."[1]

A more complete account of these discoveries will be prepared for the Fourth Volume of the Collections of the Ayrshire
and Wigtownshire Archæological Association.

[1] See article on Inscribed Stones at Kirkmadrine, in the parish of
Stonykirk, county of Wigtown, by Dr. Arthur Mitchell.—*Proc. Soc.
Antiq. Scot.* vol. ix. p. 568.   Also, *Scotland in Early Christian Times*
(Second Series), p. 252.

INDEX.

THE END.

Edinburgh University Press:
THOMAS AND ARCHIBALD CONSTABLE, PRINTERS TO HER MAJESTY.